OVERCOMING BARRIERS TO DEPLOYMENT OF PLUG-IN ELECTRIC VEHICLES

Committee on Overcoming Barriers to Electric-Vehicle Deployment

Board on Energy and Environmental Systems

Division on Engineering and Physical Sciences

and

Transportation Research Board

NATIONAL RESEARCH COUNCIL
OF THE NATIONAL ACADEMIES

THE NATIONAL ACADEMIES PRESS
Washington, D.C.
www.nap.edu

THE NATIONAL ACADEMIES PRESS **500 Fifth Street, NW** **Washington, DC 20001**

NOTICE: The project that is the subject of this report was approved by the Governing Board of the National Research Council, whose members are drawn from the councils of the National Academy of Sciences, the National Academy of Engineering, and the Institute of Medicine. The members of the committee responsible for the report were chosen for their special competences and with regard for appropriate balance.

This project was supported by Contract DE-EE0004436 between the National Academy of Sciences and the U.S. Department of Energy. Any opinions, findings, conclusions, or recommendations expressed in this publication are those of the authors and do not necessarily reflect the view of the organizations or agencies that provided support for this project.

International Standard Book Number-13: 978-0-309-37217-6
International Standard Book Number-10: 0-309-37217-8

Library of Congress Control Number: 2015939639

Additional copies of this report are available for sale from the National Academies Press, 500 Fifth Street, NW, Keck 360, Washington, DC 20001; (800) 624-6242 or (202) 334-3313; http://www.nap.edu.

Printed in the United States of America.

THE NATIONAL ACADEMIES

Advisers to the Nation on Science, Engineering, and Medicine

The **National Academy of Sciences** is a private, nonprofit, self-perpetuating society of distinguished scholars engaged in scientific and engineering research, dedicated to the furtherance of science and technology and to their use for the general welfare. Upon the authority of the charter granted to it by the Congress in 1863, the Academy has a mandate that requires it to advise the federal government on scientific and technical matters. Dr. Ralph J. Cicerone is president of the National Academy of Sciences.

The **National Academy of Engineering** was established in 1964, under the charter of the National Academy of Sciences, as a parallel organization of outstanding engineers. It is autonomous in its administration and in the selection of its members, sharing with the National Academy of Sciences the responsibility for advising the federal government. The National Academy of Engineering also sponsors engineering programs aimed at meeting national needs, encourages education and research, and recognizes the superior achievements of engineers. Dr. C. D. Mote, Jr., is president of the National Academy of Engineering.

The **Institute of Medicine** was established in 1970 by the National Academy of Sciences to secure the services of eminent members of appropriate professions in the examination of policy matters pertaining to the health of the public. The Institute acts under the responsibility given to the National Academy of Sciences by its congressional charter to be an adviser to the federal government and, upon its own initiative, to identify issues of medical care, research, and education. Dr. Victor J. Dzau is president of the Institute of Medicine.

The **National Research Council** was organized by the National Academy of Sciences in 1916 to associate the broad community of science and technology with the Academy's purposes of furthering knowledge and advising the federal government. Functioning in accordance with general policies determined by the Academy, the Council has become the principal operating agency of both the National Academy of Sciences and the National Academy of Engineering in providing services to the government, the public, and the scientific and engineering communities. The Council is administered jointly by both Academies and the Institute of Medicine. Dr. Ralph J. Cicerone and Dr. C. D. Mote, Jr., are chair and vice chair, respectively, of the National Research Council.

www.national-academies.org

COMMITTEE ON OVERCOMING BARRIERS TO ELECTRIC-VEHICLE DEPLOYMENT

JOHN G. KASSAKIAN, *Chair*, NAE,[1] Massachusetts Institute of Technology
DAVID BODDE, Clemson University, Clemson, South Carolina
JEFF DOYLE, D'Artagnan Consulting, Olympia, Washington
GERALD GABRIELSE, NAS,[2] Harvard University, Cambridge, Massachusetts
KELLY SIMS GALLAGHER, Tufts University, Medford, Massachusetts (until June 2014)
ROLAND HWANG, Natural Resources Defense Council, San Francisco, California
PETER ISARD, Consultant, Washington, D.C.
LINOS JACOVIDES, NAE, Michigan State University, East Lansing
ULRIC KWAN, IBM Global Business Services, Palo Alto, California
REBECCA LINDLAND, King Abdullah Petroleum Studies and Research Center, Riyadh, Saudi Arabia
RALPH MASIELLO, NAE, DNVGL, Inc., Chalfont, Pennsylvania
JAKKI MOHR, University of Montana, Missoula
MELISSA SCHILLING, New York University, Stern School of Business, New York
RICHARD TABORS, Across the Charles, Cambridge, Massachusetts
THOMAS TURRENTINE, University of California, Davis

Staff

ELLEN K. MANTUS, Project Codirector
K. JOHN HOLMES, Project Codirector
JAMES ZUCCHETTO, Board Director
JOSEPH MORRIS, Senior Program Officer
LIZ FIKRE, Senior Editor
MICHELLE SCHWALBE, Program Officer
ELIZABETH ZEITLER, Associate Program Officer
IVORY CLARKE, Senior Program Assistant
LINDA CASOLA, Senior Program Assistant

---

[1] NAE, National Academy of Engineering.

[2] NAS, National Academy of Sciences.

BOARD ON ENERGY AND ENVIRONMENTAL SYSTEMS

ANDREW BROWN, JR., *Chair*, NAE,[1] Delphi Corporation, Troy, Michigan
DAVID T. ALLEN, University of Texas, Austin
W. TERRY BOSTON, NAE, PJM Interconnection, LLC, Audubon, Pennsylvania
WILLAM BRINKMAN, NAS,[2] Princeton University, Princeton, New Jersey
EMILY CARTER, NAS, Princeton University, Princeton, New Jersey
CHRISTINE EHLIG-ECONOMIDES, NAE, Texas A&M University, College Station
NARAIN HINGORANI, NAE, Independent Consultant, San Mateo, California
DEBBIE NIEMEIER, University of California, Davis
MARGO OGE, McLean, Virginia
MICHAEL OPPENHEIMER, Princeton University, Princeton, New Jersey
JACKALYNE PFANNENSTIEL, Independent Consultant, Piedmont, California
DAN REICHER, Stanford University, Stanford, California
BERNARD ROBERTSON, NAE, Daimler-Chrysler (retired), Bloomfield Hills, Michigan
DOROTHY ROBYN, Washington, D.C.
GARY ROGERS, Roush Industries, Livonia, Michigan
ALISON SILVERSTEIN, Consultant, Pflugerville, Texas
MARK THIEMENS, NAS, University of California, San Diego
ADRIAN ZACCARIA, NAE, Bechtel Group, Inc. (retired), Frederick, Maryland
MARY LOU ZOBACK, NAS, Stanford University, Stanford, California

Staff

JAMES ZUCCHETTO, Board Director
DANA CAINES, Financial Associate
ALAN CRANE, Senior Scientist
K. JOHN HOLMES, Senior Program Officer/Associate Director
MARTIN OFFUTT, Senior Program Officer
ELIZABETH ZEITLER, Associate Program Officer
LANITA JONES, Administrative Coordinator
LINDA CASOLA, Senior Program Assistant
ELIZABETH EULLER, Program Assistant
JONATHAN YANGER, Research Associate

---

[1] NAE, National Academy of Engineering.

[2] NAS, National Academy of Sciences.

TRANSPORTATION RESEARCH BOARD 2014 EXECUTIVE COMMITTEE[1]

[1] Membership as of October 2014.

Preface

The plug-in electric vehicle (PEV) holds much promise—from reducing dependence on imported petroleum to decreasing greenhouse gas emissions to improving urban air quality. However, there are many barriers to its mainstream adoption regardless of incentives and enticing promises to solve difficult problems. Such vehicles have some limitations owing to current battery technology, such as restricted electric driving range and the long times required for battery charging. Furthermore, they cost more than conventional vehicles and require an infrastructure for charging the battery. Given those concerns, the U.S. Congress asked the Department of Energy to commission a study by the National Research Council (NRC) that would investigate the barriers and recommend ways to overcome them.

In this final comprehensive report, the Committee on Overcoming Barriers to Electric-Vehicle Deployment first discusses the current characteristics of PEVs and charging technologies. It then briefly reviews the market-development process, presents consumer demographics and attitudes toward PEVs, and discusses the implications of that information and other factors on PEV adoption and diffusion. The committee next explores how federal, state, and local governments and their various administrative arms can be more supportive and implement policies to sustain beneficial strategies for PEV deployment. It then provides an in-depth discussion of the PEV charging-infrastructure needs and evaluates the implications of PEV deployment on the electricity sector. Finally, the committee discusses incentives for adopting PEVs.

The current report has been reviewed in draft form by persons chosen for their diverse perspectives and technical expertise in accordance with procedures approved by the NRC Report Review Committee. The purpose of the independent review is to provide candid and critical comments that will assist the institution in making its published report as sound as possible and to ensure that the report meets institutional standards of objectivity, evidence, and responsiveness to the study charge. The review comments and draft manuscript remain confidential to protect the integrity of the deliberative process. We thank the following people for their review of this report:

Ron Adner, Dartmouth College,
William F. Brinkman, NAS, Princeton University,
Yet-Ming Chiang, NAE, Massachusetts Institute of Technology,
George Eads, Charles River Associates,
Gregory A. Franklin, University of Alabama at Birmingham,
John D. Graham, Indiana University,
Christopher T. Hendrickson, NAE, Carnegie Mellon University,
Jeremy J. Michalek, Carnegie Mellon University,
John O'Dell, Edmunds.com,
Margo Tsirigotis Oge, U.S. Environmental Protection Agency (retired),
Karl Popham, Austin Energy, and
Mike Tamor, Ford Motor Company.

Although the reviewers listed above have provided many constructive comments and suggestions, they were not asked to endorse the conclusions or recommendations, nor did they see the final draft of the report before its release. The review of the report was overseen by the review coordinator, Maxine Savitz, NAE, Honeywell Inc. (retired), and the review monitor, M. Granger Morgan, NAS, Carnegie Mellon University. Appointed by NRC, they were responsible for making certain that an independent examination of the report was carried out in accordance with institutional procedures and that all review comments were carefully considered. Responsibility for the final content of the report rests entirely with the committee and the institution. The committee gratefully acknowledges the following for their presentations during open sessions of the committee meetings:

Ali Ahmed, Cisco Systems, Inc.,
Marcus Alexander, Electric Power Research Institute,
Menahem Anderman, Advanced Automotive Batteries,
Greg Brown, Serra Chevrolet,
Allison Carr, Houston-Galveston Area Clean Cities Coalition,
William P. Chernicoff, Toyota Motors North America, Inc.,
Mike Cully, Car2Go,
Tammy Darvish, DARCARS Automotive Group,
Patrick B. Davis, U.S. Department of Energy,
Katie Drye, Advanced Energy,
Rick Durst, Portland General Electric,

Alexander Edwards, Strategic Vision,
James Francfort, Idaho National Laboratory,
Linda Gaines, Argonne National Laboratory,
Camron Gorguinpour, U.S. Department of Defense,
David Greene, Oak Ridge National Laboratory,
Doug Greenhaus, National Automobile Dealers Association,
Britta K. Gross, General Motors,
Jonna Hamilton, Electrification Coalition,
Steve Hanson, Frito-Lay,
Jack Hidary, Hertz,
John H. Holmes, San Diego Gas and Electric,
Dana Jennings, Lynda.com, Inc.,
Donald Karner, ECOtality North America,
Elise Keddie, California Air Resources Board,
Ed Kim, AutoPacific,
Neil Kopit, Criswell Automotive,
Michael Krauthamer, eVgo,
Richard Lowenthal, ChargePoint,
Brewster McCracken, Pecan Street Inc.,
John Miller, JNJ Miller plc,
Russ Musgrove, FedEx Express,
Michael Nicholas, Institute of Transportation Studies, University of California, Davis,
Nick Nigro, Center for Climate and Energy Solutions,
Sarah Olexsak, U.S. Department of Energy,
John Rhow, Kleiner Perkins,
Paul Scott, Downtown Los Angeles Nissan,
Chuck Shulock, Shulock Consulting,
Lee Slezak, U.S. Department of Energy,
John Smart, Idaho National Laboratory,
Suresh Sriramulu, TIAX LLC,
Mark Sylvia, Massachusetts Department of Energy Resources,
Mike Tamor, Ford Motor Company,
Joseph Thompson, Nissan,
Chris Travell, Maritz Research,
Jacob Ward, U.S. Department of Energy,
Jason Wolf, Better Place, and
Tracy Woodard, Nissan.

The committee also wishes to express its gratitude to Tomohisa Maruyama, Ministry of Economy, Trade, and Industry, Tokyo, Japan, and Sumiyo Hirano, Next Generation Vehicle Promotion Center, Tokyo, Japan, for arranging an informative visit to Japan and accompanying the members as they traveled through Japan. The committee also wishes to thank the following for providing valuable information and extending hospitality to the committee during its visits to Germany, Japan, The Netherlands, and Texas:

Austin Energy, Austin, Texas,
Berlin Agency for Electric Mobility (eMO), Berlin, Germany,
Charging Network Development Organization, Tokyo, Japan,
Climate Change Policy Headquarters, City of Yokohama,
Federal Government Joint Unit for Electric Mobility (GGEMO), Berlin, Germany,
German Institute for Transportation Research (DLR), Berlin, Germany,
Innovation Centre for Mobility and Societal Change, Berlin, Germany,
Japan Charge Network, Co., Kanagawa, Japan,
Kanagawa Prefectural Government, Kanagawa, Japan,
Kyoto Prefectural Government, Kyoto, Japan,
Ministry of Economy, Trade, and Industry, Tokyo, Japan,
Ministry of Infrastructure and the Environment and Netherlands School of Public Administration, The Hague, The Netherlands,
MRA-Elektrisch, Amsterdam, The Netherlands,
Nissan Motor Co., Yokohama, Japan,
NRG eVgo, Houston, Texas,
Okayama Vehicle Engineering Center, Okayama, Japan,
Osaka Prefectural Government, Osaka, Japan,
Pecan Street Research Institute, Austin, Texas,
Technical University of Eindhoven and BrabantStad, Eindhoven, The Netherlands,
Tesla, The Netherlands,
Tokyo Electric Power Company, Kanagawa, Japan,
Urban Development Group, City of Rotterdam, The Netherlands, and
Vattenfall, Berlin, Germany.

The committee is also grateful for the assistance of the NRC staff in preparing this report. Staff members who contributed to the effort are Ellen Mantus and K. John Holmes, Project Codirectors; James Zucchetto, Director of the Board on Energy and Environmental Systems; Joseph Morris, Senior Program Officer for the TRB; Liz Fikre, senior editor; Michelle Schwalbe, Program Officer; Elizabeth Zeitler, Associate Program Officer, and Ivory Clarke and Linda Casola, Senior Program Assistants.

I especially thank the members of the committee for their efforts throughout the development of this report.

John G. Kassakian, *Chair*
Committee on Overcoming Barriers to Electric-Vehicle Deployment

Contents

SUMMARY 1

1 **INTRODUCTION** **8**
Historical and Policy Context, 8
The Plug-in Electric Vehicle and Current Sales, 9
Plug-in Electric Vehicles: Benefits and Trade-offs, 12
The Committee and Its Task, 13
The Committee's Approach to Its Task, 16
Organization of This Report, 16
References, 17

2 **PLUG-IN ELECTRIC VEHICLES AND CHARGING TECHNOLOGIES** **19**
Types of Plug-in Electric Vehicles, 19
High-Energy Batteries, 22
Relative Costs of Plug-in Electric and ICE Vehicles, 25
Vehicle Charging and Charging Options, 29
References, 34

3 **UNDERSTANDING THE CUSTOMER PURCHASE AND MARKET DEVELOPMENT PROCESS FOR PLUG-IN ELECTRIC VEHICLES** **37**
Understanding and Predicting the Adoption of New Technologies, 37
Demographics and Implications for Adoption and Diffusion of Vehicles, 39
The Mainstream Consumer and Possible Barriers to Their Adoption of Plug-in Electric Vehicles, 46
Vehicle Dealerships: A Potential Source of Information? 51
Strategies to Overcome Barriers to Deployment of Plug-in Electric Vehicles, 53
Federal Government Efforts to Familiarize Consumers with Plug-in Electric Vehicles: Clean Cities Coalition, 57
Fleet Purchases, 58
References, 60

4 **GOVERNMENT SUPPORT FOR DEPLOYMENT OF PLUG-IN ELECTRIC VEHICLES** **65**
Federal Government Research Funding to Support Deployment of Plug-in Electric Vehicles, 65
Institutional Support for Promoting Plug-in Electric Vehicle Readiness, 66
Transportation Taxation and Financing Issues Related to Plug-in Electric Vehicles, 67
Streamlining Codes, Permits, and Regulations, 74
Ancillary Institutional Issues Related to Support for Plug-in Electric Vehicles, 75
References, 78

5 **CHARGING INFRASTRUCTURE FOR PLUG-IN ELECTRIC VEHICLES** **82**
Charging Infrastructure and Effects on Deployment of Plug-in Electric Vehicles and on Electric Vehicle Miles Traveled, 82
Models for Infrastructure Deployment, 90
References, 95

6 **IMPLICATIONS OF PLUG-IN ELECTRIC VEHICLES FOR THE ELECTRICITY SECTOR**........................**98**
The Physical and Economic Structure of the Electricity Sector, 99
Generation and Transmission, 100
Physical Constraints in the Distribution Infrastructure, 100
Potential Economic Constraints or Impediments within the Delivery System, 102
Electricity Sector Regulatory Issues for Operating a Public Charging Station, 105
The Utility of the Future, 107
References, 107

7 **INCENTIVES FOR THE DEPLOYMENT OF PLUG-IN ELECTRIC VEHICLES**..................................**109**
Vehicle Price and Cost of Ownership, 109
Price and Cost Competitiveness of Plug-in Electric Vehicles, 110
Possibilities for Declines in Production Costs for Plug-in Electric Vehicles, 112
Incentives, 113
Price of Conventional Transportation Fuels as an Incentive or a Disincentive for the Adoption of Plug-in Electric Vehicles, 119
Past Incentives on Other Alternative Vehicles and Fuels, 120
Recommendations, 123
References, 123

APPENDIXES

A **BIOGRAPHICAL INFORMATION ON THE COMMITTEE ON OVERCOMING BARRIERS TO ELECTRIC-VEHICLE DEPLOYMENT**..........**129**
B **MEETINGS AND PRESENTATIONS**..........**133**
C **INTERNATIONAL INCENTIVES**..........**136**

BOXES, FIGURES, AND TABLES

BOXES

1-1 Statement of Task, 14
3-1 Calculating Electricity or Fuel Costs for Plug-in Electric and Other Vehicles, 49
5-1 Some Hypothetical Economics for Providers of Public Charging, 95
7-1 Derivation of Petroleum Equivalent for a Battery Electric Vehicle, 115
7-2 Financial Incentives, 117

FIGURES

1-1 U.S. BEV monthly sales data from 2010 to 2014, 10
1-2 U.S. PHEV monthly sales data from 2010 to 2014, 11
1-3 World PEV sales in 2012, 2013, and 2014, 11
1-4 The rate of PEV market growth in its first 34 months superimposed on the rate of HEV market growth during its first 34 months, 12
1-5 Projected annual light-duty PEV sales as a percentage of total light-duty vehicle sales, 13
2-1 The volume energy density and the mass energy density for various battery types, 23
2-2 Representation of a lithium-ion battery that shows lithium ions traveling between the anode and the cathode and electrons traveling through the external circuit to produce an electric current, 24
2-3 Effect of ambient temperature on battery capacity on a 20 kWh battery in a PHEV, 26
2-4 Change in the sales price of NiMH, Li-ion, and NiCd battery cells from 1999 to 2012, 28
2-5 For AC level 1, a vehicle is plugged into a single-phase 120 V electric socket through a portable safety device called an electric vehicle supply equipment (EVSE), 30
2-6 The SAE J1772 plug that connects all PEVs to AC level 1 and level 2 is an agreed-on universal standard for 120 V and 240 V ac charging, 30
2-7 For AC level 2 charging, a vehicle is plugged into a split-phase 240 V electric circuit like those used by electric dryers, stoves, and large air conditioners through a wall- or post-mounted safety device called an electric vehicle supply equipment (EVSE), 31

2-8 Four plugs and control protocols are now being used for DC fast charging, 32
2-9 DC fast charging a Nissan Leaf, 32
2-10 As of February 2015, Tesla had installed 190 units in the United States, 33
3-1 Years needed for fastest growing consumer technologies to achieve penetration (0-50 percent or 51-80 percent), 38
3-2 Distribution of adopter categories, 40
3-3 Women's rate of participation in the markets for all vehicles and for PEVs, 42
3-4 Projected 2014 light-duty PEV volume in the 100 largest MSAs, 43
3-5 Worldwide growth of car sharing in terms of vehicles and members, 45
3-6 Clean Cities coalitions funded for community-readiness and planning for PEVs and PEV charging infrastructure, 58
3-7 Fleet sales for passenger vehicles for 2012 by fleet purchase agency, 59
4-1 Corporate Average Fuel Economy requirements by year, 68
4-2 Sources of revenue for the federal Highway Trust Fund, FY 2010, 69
4-3 U.S. annual light-duty fuel consumption and VMT, 70
4-4 Annual transportation-related taxes paid by Washington state drivers, 71
4-5 Historic and forecast gasoline-tax revenue for Washington state, FY 1990 to FY 2040, 73
4-6 PEV-specific measures for transportation funding, 74
5-1 PEV charging infrastructure categories, ranked by their likely importance to PEV deployment, with the most important, home charging, on the bottom, and the least important, interstate DC fast charging, at the top, 83
5-2 Vehicle locations throughout the week on the basis of data from the 2001 National Household Travel Survey, 84
6-1 U.S. electricity demand growth, 1950-2040, 99
6-2 Schematic of U.S. electric power delivery system, 99
6-3 Hourly demand for electricity at a substation in a residential distribution system, 101
6-4 Residential charging behavior in NES and PG&E service territories, as measured in the EV Project, 104
6-5 States that have regulations regarding who can own or operate a PEV charging station, 106
7-1 Japan's clean energy vehicles promotion program, 119
7-2 U.S. HEV and PEV sales overlaid with U.S. gasoline prices, 120

TABLES

S-1 Four Classes of Plug-in Electric Vehicles, 2
S-2 Effects of Charging Infrastructure by PEV Class and Entities Motivated to Install Infrastructure Categories, 5
2-1 Definitions and Examples of the Four Types of Plug-in Electric Vehicles, 20
2-2 Properties of Lithium-Ion Batteries in Four Plug-in Electric Vehicles on the U.S. Market, 24
2-3 Estimates of Dollars per Kilowatt-hour for a 25 kWh Battery, 27
2-4 Summary of Estimated Costs of Total Energy from Various Sources (2013 U.S. $/kWh), 28
3-1 Categories and Descriptions of Adopters, 40
3-2 Comparison of New BEV Buyers, PHEV Buyers, and ICE-Vehicle Buyers, 41
3-3 Comparison of All New-Vehicle Buyers to Buyers of Specific Plug-in Electric Vehicles, 41
3-4 Factors That Affect Adoption and Diffusion of Innovation, 46
3-5 Consumer Questions Related to Plug-in Electric Vehicle (PEV) Ownership, 51
3-6 Ratings of Dealer Knowledge about Various Topics, 53
3-7 Websites with Information on Plug-in Electric Vehicles, 55
3-8 Information Resources for Fleet Managers, 59
4-1 Factors Determining PEV Readiness and Organizations Involved, 67
4-2 Comparison of Unrealized Revenue from Battery Electric Vehicles and Plug-in Hybrid Electric Vehicles, 72
4-3 Types of Equity and Examples in the Transportation Tax System, 72
4-4 Variation in Residential Electric Permit Fees by City or State, 75
5-1 Effect of Charging-Infrastructure Categories on Mainstream PEV Owners by PEV Class, 85
5-2 Charging Patterns for Nissan Leafs and Chevrolet Volts, 87
5-3 Entities That Might Have an Incentive to Install Each Charging Infrastructure Category, 91
5-4 Costs of Installing Public DC Fast-Charging Stations for the West Coast Electric Highway Project, 93
6-1 Definitions, Advantages, and Disadvantages of Various Types of Electric Rates, 103
7-1 MSRPs and 5-year Cumulative Cost of Ownership for Selected Plug-in Electric Vehicles and Comparative Vehicles (dollars), 111
7-2 Incentives for Plug-in Electric Vehicles (PEVs) by Country and State, 118

Summary

The plug-in electric vehicle (PEV) has a long history. In 1900, 28 percent of the passenger cars sold in the United States were electric, and about one-third of the cars on the road in New York City, Boston, and Chicago were electric. Then, however, mass production of an inexpensive gasoline-powered vehicle, invention of the electric starter for the gasoline vehicle, a supply of affordable gasoline, and development of the national highway system, which allowed long-distance travel, led to the demise of those first PEVs. In the 1970s and 1990s, interest in PEVs resurfaced, but the vehicles simply could not compete with gasoline-powered ones. In the last few years, interest in PEVs has been reignited because of advances in battery and other technologies, new federal standards for carbon-dioxide emissions and fuel economy, state zero-emission-vehicle requirements, and the current administration's goal of putting millions of alternative-fuel vehicles on the road. People are also beginning to recognize the advantages of PEVs over conventional vehicles, such as lower operating costs, smoother operation, and better acceleration; the ability to fuel up at home; and zero tailpipe emissions when the vehicle operates solely on its battery. There are, however, barriers to PEV deployment, including the vehicle cost, the short all-electric driving range, the long battery-charging time, uncertainties about battery life, the few choices of vehicle models, and the need for a charging infrastructure to support PEVs whether at home, at work, or in a public space. Moreover, many people are still not aware of or do not fully understand the new technology. Given those recognized barriers to PEV deployment, Congress asked the Department of Energy (DOE) to commission a study by the National Academies to address market barriers that are slowing the purchase of PEVs and hindering the deployment of supporting infrastructure.[1] Accordingly, the National Research Council (NRC), an arm of the National Academies, appointed the Committee on Overcoming Barriers to Electric-Vehicle Deployment, which prepared this report.

[1] See Consolidated Appropriations Act, 2012, P.L. 112-74, H. Rept. 112-331 (H.Rept. 112-118).

THE COMMITTEE'S TASK

The committee's analysis was to be provided in two reports—a short interim report and a final comprehensive report. The committee's interim report, released in May 2013, provided an initial discussion of infrastructure needs for PEVs, barriers to deploying the infrastructure, and possible roles for the federal government in overcoming the barriers. It did not offer any recommendations because the committee was still in the early stages of gathering data. The current report is the committee's final comprehensive report that addresses its full statement of task, which can be found in Chapter 1.

This report focuses on light-duty vehicles (passenger cars and light-duty trucks) in the United States and restricts its discussion to PEVs, which include battery electric vehicles (BEVs) and plug-in hybrid electric vehicles (PHEVs).[2] The common feature of these vehicles is that they can charge their batteries by plugging into the electric grid. The distinction between them is that BEVs operate solely on electricity stored in the battery (there is no other energy source), and PHEVs have an internal-combustion engine (ICE) that can supplement the electric power train or charge the battery during a trip. PHEVs can use engines powered by various fuels, but this report focuses on those powered by gasoline because they are the ones currently available in the United States.

The premise of the committee's task is that there is a benefit to the United States if a higher fraction of miles is fueled by electricity rather than by petroleum. Two reasons for this benefit are commonly assumed. First, a higher fraction of miles fueled by electricity would reduce the U.S. dependence on petroleum. Second, a higher fraction of miles fueled by electricity would reduce carbon dioxide and other air pollutants emitted into the atmosphere. The committee

[2] BEVs and PHEVs need to be distinguished from conventional hybrid electric vehicles (HEVs), such as the Toyota Prius that was introduced in the late 1990s. HEVs do not plug into the electric grid but power their batteries from regenerative braking and an internal-combustion engine. They are not included in the PEV category and are not considered further in this report unless to make a comparison on some issue.

was not asked to research or evaluate the premise, but it did consider whether the premise was valid now and into the future and asked if any recent developments might call the premise into question.

First, a PEV uses no petroleum when it runs on electricity. Furthermore, the electricity that fuels the vehicle is generated using essentially no petroleum; in 2013, less than 0.7 percent of the U.S. grid electricity was produced from petroleum. Thus, PEVs advance the long-term objective of U.S. energy independence and security. Second, on average, a PEV fueled by electricity is now responsible for less greenhouse gases (GHGs) per mile than an ICE vehicle[3] or a hybrid electric vehicle (HEV). PEVs will make further reductions in GHG emissions as the U.S. electric grid changes to lower carbon sources for its electricity. Therefore, the committee concludes that the premise for the task—that there is an advantage to the United States if a higher fraction of miles driven here are fueled by electricity from the U.S. electric grid—is valid now and becomes even more valid each year that the United States continues to reduce the GHGs that it produces in generating electricity. A more detailed discussion of the committee's analysis of the near-term and long-term impacts of PEV deployment on petroleum consumption and GHG emissions is provided in Chapter 1 of this report.

Recommendation: As the United States encourages the adoption of PEVs, it should continue to pursue in parallel the production of U.S. electricity from increasingly lower carbon sources.

[3] For this report, *ICE vehicle* or *conventional vehicle* refers to a light-duty vehicle that obtains all of its propulsion from an internal-combustion engine.

PLUG-IN ELECTRIC VEHICLES AND CHARGING TECHNOLOGIES

Today, there are several makes and models of PEVs on the market, and PEV sales reached about 0.76 percent of the light-duty sales in the United States by the close of 2014. Because the obstacles to consumer adoption and the charging infrastructure requirements depend on PEV type, the committee used the all-electric range (AER) of the vehicles to distinguish four PEV classes (see Table S-1). Several important points regarding the PEV classes should be highlighted. First, the Tesla Model S clearly demonstrates the possibility of producing a long-range BEV that has been recognized as a high-performing vehicle. Second, limited-range BEVs are the *only* type of PEV that have a substantial range limitation. Although they are not practical for trips that would require more than one fast charge given the substantial refueling time required, their ranges are more than sufficient for the average daily travel needs of the majority of U.S. drivers. Third, the range-extended PHEV has a total range that is comparable to that of a conventional vehicle because of the onboard ICE, and the typical AER is comparable to or larger than the average U.S. daily travel distance. The fraction of miles traveled by electricity depends on how willing and able a driver is to recharge the battery during a trip longer than the AER. Fourth, minimal PHEVs with AERs much shorter than the average daily driving distance in the United States are essentially HEVs.

There are three options for charging the high-energy batteries in PEVs.[4] First, AC level 1 uses a 120 V circuit and provides about 4-5 miles of electric range per hour of

[4] A fourth option might be considered wireless charging, but this option is not widely used today.

TABLE S-1 Four Classes of Plug-in Electric Vehicles

PEV Class	Description	Example (Range[a])
Long-range BEV	Can travel hundreds of miles on a single battery charge and then be refueled in a time that is much shorter than the additional driving time that the refueling allows.	2014 Tesla Model S (AER = 265 miles)
Limited-range BEV	Is made more affordable than the long-range BEV by reducing the size of the high-energy battery. Its limited range can more than suffice for many commuters, but it is impractical for long trips.	2014 Nissan Leaf (AER = 84 miles) 2014 Ford Focus Electric (AER = 76 miles)
Range-extended PHEV	Typically, operates as a zero-emission vehicle until its battery is depleted, whereupon an ICE turns on to extend its range.	2014 Chevrolet Volt (AER = 38 miles; total range = 380 miles)
Minimal PHEV	Its small battery can be charged from the grid, but its AER is much less than the average daily U.S. driving distance.	2014 Toyota Plug-in Prius (AER = 6-11 miles; total range = 540 miles)

[a] The AERs noted are average values estimated by the U.S. Environmental Protection Agency. Total ranges are provided for PHEVs; the AER is the total range for BEVs.

NOTE: AER, all-electric range; BEV, battery electric vehicle; ICE, internal-combustion engine; PEV, plug-in electric vehicle; PHEV, plug-in hybrid electric vehicle.

charging. It is considered too slow to be the primary charging method for fully depleted batteries of PEVs that have large batteries because charging times would be longer than the time a vehicle is normally parked at home or the workplace. Second, AC level 2 uses a 240 V, split-phase ac circuit like those used by electric dryers, electric stoves or ovens, and large air conditioners; it provides about 10-20 miles of electric range per hour of charging depending on how much current the vehicle is allowed to draw. Third, DC fast charging is an option available only to BEVs today and uses high-voltage circuits to charge the battery much more rapidly. DC fast charging is generally not an option for residential charging given the high-power circuits required. In the United States, there is one standard plug for the AC level 1 and AC level 2 chargers, but there are at least three incompatible plugs and communication protocols being used for DC fast charging. Plug and protocol incompatibility is a barrier to PEV adoption insofar as it prevents all PEVs from being able to charge at any fast-charging station.

Recommendation: The federal government and proactive states should use their incentives and regulatory powers to (1) eliminate the proliferation of plugs and communication protocols for DC fast chargers and (2) ensure that all PEV drivers can charge their vehicles and pay at all public charging stations using a universally accepted payment method just as any ICE vehicle can be fueled at any gasoline station.

UNDERSTANDING THE MARKET DEVELOPMENT AND CUSTOMER PURCHASE PROCESS FOR PLUG-IN ELECTRIC VEHICLES

Developers of new technologies, such as PEVs, face challenges in developing a market and motivating consumers to purchase or use their products. Incumbent technologies—in this case, ICE vehicles—can be difficult to unseat; they have years of production and design experience, which make their production costs lower than those of emerging technologies and thus more affordable. The necessary infrastructure, including the ubiquitous presence of gasoline and service stations across the United States, is well-developed. Consumers know the attributes and features to compare to evaluate their ICE-vehicle choices, and they are accustomed to buying, driving, and fueling these vehicles. Indeed, one of the main challenges to the success of the PEV market is that people are so accustomed to ICE vehicles.

Accordingly, adoption and diffusion of PEVs is likely to be a long-term, complex process. Even modest market penetration could take many years. Furthermore, market penetration rates will likely be a function not only of the product itself but also of the entire industry ecosystem. Hence, product technologies (such as low-cost batteries), downstream infrastructure (such as dealers and repair facilities), and complementary infrastructure (such as charging stations) will need to be developed simultaneously.

One strategy for dealing with market complexity has been to identify a narrow market segment for which the new technology offers a compelling reason to buy. Offering a compelling value proposition specifically targeted to meet the needs of a narrow market segment rather than the broad mass market gives the technology a greater chance to dominate in that key market segment. Then, the momentum gained in the initial market segment can be used more efficiently and effectively to drive sales in related, adjacent segments. That approach appears reasonable for PEVs because the PEV market has been characterized by strong regional patterns that reflect such attributes as expensive gasoline; favorable demographics, values, and lifestyles; a regulatory environment favorable to PEVs; and an existing or at least readily deployable infrastructure.

The purchase of a new vehicle is typically a lengthy process that often involves substantial research and is strongly affected by consumer perceptions. In evaluating the purchase process for PEVs specifically, the committee identified several barriers—in addition to the cost differences between PEVs and ICE vehicles—that affect consumer perceptions and their decision process and ultimately (negatively) their purchase decisions. The barriers include the limited variety of PEVs available; misunderstandings concerning the range of the various PEVs; difficulties in understanding electricity consumption, calculating fuel costs, and determining charging infrastructure needs; complexities of installing home charging; difficulties in determining the greenness of the vehicle; lack of information on incentives; and lack of knowledge of unique PEV benefits. Collectively, the identified barriers indicate that consumer awareness and knowledge of PEV offerings, incentives, and features are not as great as needed to make fully informed decisions about whether to purchase a PEV. Furthermore, many factors contribute to consumer uncertainty and doubt about the viability of PEVs and create a perceptual hurdle that negatively affects PEV purchases. Together, the barriers emphasize the need for better consumer information and education that can answer all their questions. Consumers have traditionally relied on dealers to provide vehicle information; however, in spite of education efforts by some manufacturers, dealer knowledge of PEVs has been uneven and often insufficient to address consumer questions and concerns. The committee does acknowledge, however, that even well-informed consumers might not buy a PEV because it does not meet some of their basic requirements for a vehicle (that is, consumer information and education cannot overcome the absence of features desired by a consumer).

Recommendation: To provide accurate consumer information and awareness, the federal government should make use of its Ad Council program, particularly in key geographic markets, to provide accurate information about federal tax credits and other incentives, the value proposition for PEV ownership, and who could usefully own a PEV.

GOVERNMENT SUPPORT FOR DEPLOYMENT OF PLUG-IN ELECTRIC VEHICLES

The federal government can play a substantive role in encouraging PEV deployment by supporting research that has the potential to remove barriers. Specifically, investment in battery research is critical for producing lower cost, higher performing batteries. Improved battery technology will lower vehicle cost, increase the all-electric range, or both, and those improvements will likely lead to increased PEV deployment. Furthermore, research is needed to understand the relationship between charging infrastructure availability and PEV adoption and use. Specifically, research should be conducted to determine how much public infrastructure is needed and where it should be sited to induce PEV adoption and to encourage PEV owners to optimize their vehicle use. That research is especially critical if the federal government is allocating resources to fund public infrastructure deployment.

Recommendation: The federal government should continue to sponsor fundamental and applied research to facilitate and expedite the development of lower cost, higher performing vehicle batteries. Stable funding is critical and should focus on improving energy density and addressing durability and safety.

Recommendation: The federal government should fund research to understand the role of public charging infrastructure (as compared with home and workplace charging) in encouraging PEV adoption and use.

The successful deployment of PEVs will involve many entities, including federal, state, and local governments. One potential barrier for PEV adoption that is solely within government control is taxation of PEVs and, in particular, taxation for the purpose of recovering the costs of maintaining, repairing, and improving roadways. In the United States, fuel taxes have been used to finance transportation budgets. Because BEVs use no gasoline and PHEVs use much less gasoline than ICE vehicles, there is the belief that PEV owners pay nothing to support transportation infrastructure and should be taxed or charged a special fee. However, PEV owners pay taxes and fees other than fuel taxes that support transportation budgets. Furthermore, the fiscal impact at the present time and likely over the next decade of not collecting fuel taxes from PEV owners is negligible, especially compared with the impact of high-mileage vehicles that are being produced to meet fuel-economy standards.

Recommendation: Federal and state governments should adopt a PEV innovation policy where PEVs remain free from special roadway or registration surcharges for a limited time to encourage their adoption.

Some federal and state permitting processes have been ill-suited for the simple installation of some PEV charging infrastructure. As a result, unnecessary permit burdens and costs have been introduced into the installation process. Because most charging will occur at home, PEV deployment could be seriously impeded if the buyers must bear high permit and installation costs and experience delay in the activation of their home chargers. Accordingly, clarity, predictability, and speed are needed in the permitting and approval process for installation of home and public charging stations.

Recommendation: Local governments should streamline permitting and adopt building codes that require new construction to be capable of supporting future charging installations.

CHARGING INFRASTRUCTURE FOR PLUG-IN ELECTRIC VEHICLES

PEV deployment and the fraction of vehicle miles fueled by electricity (eVMT) critically depend on the charging infrastructure. For its analysis, the committee categorized the infrastructure by location (home, workplace, intracity, intercity, and interstate) and power (AC level 1, AC level 2, and DC fast charging), evaluated it from the perspective of the PEV classes defined in Table S-1, and determined which entities might have a motivation to install which category of charging infrastructure. The results of the committee's analysis are summarized in Table S-2. The table reflects the relative importance of each infrastructure category as assessed by the committee, with home listed first (most important) and interstate listed last (least important).

Several points should be made for the various infrastructure categories. First, home charging is a virtual necessity for all PEV classes given that the vehicle is typically parked at a residence for the longest portion of the day. Accordingly, the home is (and will likely remain) the most important location for charging infrastructure, and homeowners who own PEVs have a clear incentive to install home charging. Residences that do not have access to a dedicated parking spot or one with access to electricity clearly have challenges to overcome to make PEV ownership practical for them.

Second, charging at workplaces offers an important opportunity to encourage PEV adoption and increase eVMT. Specifically, it could double the daily travel distance that is fueled by electricity if combined with home charging and could in principle make possible the use of limited-range BEVs when no home charging is available. Some businesses appear to be motivated to provide workplace charging as a means to attract and retain employees or to brand the company with a green image. However, one concern is that utilities could impose demand charges if the businesses exceed their maximum power-demand thresholds; such charges could be substantial. Another concern is the IRS requirement for businesses to assess the value of the charging and report it as imputed income.

Recommendation: Local governments should engage with and encourage workplaces to consider investments in charging infrastructure and provide information about best practices.

TABLE S-2 Effects of Charging Infrastructure by PEV Class and Entities Motivated to Install Infrastructure Categories[a]

Infrastructure Category[b]	PEV Class	Effect of Infrastructure on Mainstream PEV Owner	Who Has an Incentive to Install?
Home AC levels 1 and 2	Long-range BEV Limited-range BEV Range-extended PHEV Minimal PHEV	Virtual necessity Virtual necessity Virtual necessity Virtual necessity	Vehicle Owner, Utility
Workplace AC levels 1 and 2	Long-range BEV Limited-range BEV Range-extended PHEV Minimal PHEV	Range extension, expands market Range extension, expands market Increases eVMT and value proposition; expands market Increases eVMT and value proposition; expands market	Business Owner, Utility
Intracity[c] AC levels 1 and 2	Long-range BEV Limited-range BEV Range-extended PHEV Minimal PHEV	Not necessary Range extension, increases confidence Increases eVMT and value proposition Increases eVMT and value proposition	Utility, Retailer, Charging Provider, Vehicle Manufacturer
Intracity[c] DC fast charge	Long-range BEV Limited-range BEV Range-extended PHEV Minimal PHEV	Not necessary Range extension, increases confidence NA – not equipped NA – not equipped	Utility, Charging Provider, Vehicle Manufacturer, Government
Intercity[c] DC fast charge	Long-range BEV Limited-range BEV Range-extended PHEV Minimal PHEV	Range extension, expands market 2 × Range extension, increases confidence NA – not equipped NA – not equipped	Vehicle Manufacturer, Government
Interstate DC fast charge	Long-range BEV Limited-range BEV Range extended PHEV Minimal PHEV	Range extension, expands market Not practical for long trips NA – not equipped NA – not equipped	Vehicle Manufacturer, Government

[a] Assumptions for analysis are that electricity costs would be cheaper than gasoline costs, that away-from-home charging would generally cost as much as or more than home charging, that people would not plan to change their mobility needs to acquire a PEV, and that there would be no disruptive changes to current PEV performance and only incremental improvements in battery capacity over time.
[b] The term *intercity* refers to travel over distances less than twice the range of limited-range BEVs, and the term *interstate* refers to travel over longer distances.
[c] It is possible that these infrastructure categories could expand the market for the various types of PEVs as appropriate, but that link is more tenuous than the cases noted in the table for other infrastructure categories.
NOTE: AC, alternating current; BEV, battery electric vehicle; DC, direct current; eVMT, electric vehicle miles traveled; NA, not applicable; PEV, plug-in electric vehicle; PHEV, plug-in hybrid electric vehicle.

Third, public charging infrastructure has the potential to provide range confidence and extend the range for limited-range BEV drivers, allow long-distance travel for long-range BEV drivers, and increase eVMT and the value proposition for PHEV drivers. However, fundamental questions that need to be answered are how much and what type of public charging infrastructure is needed and where should it be located? Furthermore, although the committee has identified several entities that might be motivated to install public charging infrastructure, it could identify only two entities—BEV manufacturers and utilities—that might have an attractive business case for absorbing the full capital costs of investments in public charging infrastructure. The government might decide that providing public charging infrastructure serves a public good when others do not have a business case or incentive to do so.

Recommendation: The federal government should refrain from additional direct investment in the installation of public charging infrastructure pending an evaluation of the relationship between the availability of public charging and PEV adoption or use.

IMPLICATIONS OF PLUG-IN ELECTRIC VEHICLES FOR THE ELECTRICITY SECTOR

An important concern raised by the public and policy makers pertains to the capability of electric utilities to provide for PEV charging. At the current time, PEV charging requirements account for about 0.02 percent of the energy produced and consumed in the continental United States. Were the PEV fleet to reach as high as 20 percent of private vehicles, the estimated impact would still be only 5 percent of today's electric production. Accordingly, PEV deployment is not constrained by the transmission system or the generation capacity. Although some capital investment in (or upgrades to) the distribution infrastructure might be required in areas where there is high, concentrated PEV deployment, PEV charging is expected to have a negligible effect on the distribution system at the anticipated rates of PEV adoption.

Thus, the constraints on PEV adoption that could arise from the electricity sector are more likely to be economic rather than physical or technical. Potential impediments to PEV adoption include (1) high electricity costs that reduce the financial benefit of PEV ownership, (2) regional differences in electricity costs that add confusion and prevent a uniform explanation of the economic benefits of PEV ownership, (3) residential electric rate structures that provide no incentive to charge the vehicle at the optimal time for the utility, and (4) high costs for commercial and industrial customers if demand charges are incurred as noted above. The committee notes that state jurisdiction over retail electricity rates constrains the federal role in directing the electricity sector to foster PEV growth.

Recommendation: To ensure that adopters of PEVs have incentives to charge vehicles at times when the cost of supplying energy is low, the federal government should propose that state regulatory commissions offer PEV owners the option of purchasing electricity under time-of-use or real-time pricing.

INCENTIVES FOR THE DEPLOYMENT OF PLUG-IN ELECTRIC VEHICLES

One of the most important issues concerning PEV deployment is determining what, if any, incentives are needed to encourage PEV adoption. Determining the need for incentives is difficult because little is yet known about the effectiveness of PEV incentive programs. However, two factors to consider are vehicle price and cost of ownership. To examine those factors, the committee considered sales and consumer survey data and compared manufacturer suggested retail prices (MSRPs) on selected PEVs, HEVs, and ICE vehicles. The committee found that although sales data and consumer survey data are difficult to interpret, they are consistent with the view that price is a barrier to some buyers but that others might be rejecting PEVs for other reasons. Comparisons of MSRPs and cumulative ownership costs that incorporate current federal tax credits provide mixed evidence on whether price is an obstacle to PEV adoption. However, in the absence of tax credits or other subsidies, comparisons at today's MSRPs would be unfavorable to PEVs.

Another factor to consider is the possibility of declines in production costs for PEVs so that manufacturers can price them attractively in comparison with conventional vehicles. The decline over time in PEV production costs, however, is likely to occur gradually, and existing quotas of federal tax credits could be exhausted for manufacturers of relatively popular PEVs before costs can be substantially reduced. Thus, the deployment of PEVs might be at risk unless the federal government extends manufacturer or consumer incentives, at least temporarily.

Regulatory requirements and incentives for manufacturers and consumers have been introduced over the past few years by states and the federal government to encourage PEV production and deployment. Most manufacturer incentives and mandates are contained in the federal Corporate Average Fuel Economy standards, the federal GHG emission standards, and state zero-emission-vehicle (ZEV) programs. Most consumer incentive programs have involved purchase incentives in the form of tax credits, tax rebates, or tax exemptions. However, states have also used ownership incentives (such as exemptions from or reductions in registration taxes or fees and vehicle inspections) and use incentives (such as exemptions from motor fuel taxes, reduced roadway taxes or tolls, and discounted or free PEV charging or parking). Some states have also offered nonfinancial incentives that allow access to restricted lanes, such as bus-only, high-occupancy-vehicle, and high-occupancy-toll lanes. Incentives have also been provided to install charging stations, the availability of which might also influence people's willingness to purchase PEVs.

On the basis of the committee's analysis, several points should be highlighted. First, existing federal and state regulatory programs for fuel-economy and emissions have been effective at stimulating manufacturers to produce some PEVs, and sale of credits from these programs between manufacturers has also provided an important incentive for PEV manufacturers to price PEVs more attractively. The committee emphasizes that the state ZEV requirements have been particularly effective at increasing PEV production and adoption. Second, the effectiveness of the federal income tax credit at motivating people to purchase PEVs would be enhanced by converting it into a rebate at the point of sale. Third, state and local governments offer a variety of financial and nonfinancial incentives, but there appears to be a lack of research to indicate which incentives might be the most effective at encouraging PEV adoption. Fourth, the many state and local incentives that differ in monetary value, restrictions, and calculation methods make it challenging to educate consumers on the incentives that are available to them and emphasize the need for a clear, up-to-date source of information for consumers. Fifth, the overall international experience appears to suggest that substantial financial incentives are effective in motivating consumers to buy PEVs.

Recommendation: Federal financial incentives to purchase PEVs should continue to be provided beyond the current production volume limit as manufacturers and consumers experiment with and learn about the new technology. The federal government should re-evaluate the case for incentives after a suitable period, such as 5 years. Its re-evaluation should consider advancements in vehicle technology and progress in reducing production costs, total costs of ownership, and emissions of PEVs, HEVs, and ICE vehicles.

Recommendation: Given the research on effectiveness of purchase incentives, the federal government should consider converting the tax credit to a point-of-sale rebate.

Recommendation: Given the sparse research on incentives other than financial purchase incentives, research should be conducted on the variety of consumer incentives that are (or have been) offered by states and local governments to determine which, if any, have proven effective in promoting PEV deployment.

CONCLUDING REMARKS

The committee provides a number of recommendations throughout this report and highlights several of the most important in the summary. However, two points should be further emphasized. First, vehicle cost is a substantial barrier to PEV deployment. As noted above and discussed in detail in Chapter 7, without the federal financial purchase incentives, PEVs are not currently cost-competitive with ICE vehicles on the basis of either purchase price or cumulative cost of ownership. Therefore, one of the most important committee recommendations is continuing the federal financial purchase incentives and re-evaluating them after a suitable period. Second, developing lower cost, better performing batteries is essential for reducing vehicle cost because it is the high-energy batteries that are primarily responsible for the cost differential between PEVs and ICE vehicles. It is therefore important that the federal government continue to fund battery research at least at current levels. Technology development to improve and lower the cost of batteries (and electric-drive technologies) for PEVs represents a technology-push strategy that complements the market-pull strategy represented by the federal financial purchase incentives that lower the barrier to market adoption. A significant body of research, however, demonstrates that having the right technology (with a compelling value proposition) is still insufficient to achieve success in the market. That technology must be complemented with a planned strategy to create market awareness and to overcome customer fear, uncertainty, and doubt about the technology.

Equally important to recognize is a recommendation that the committee does not make. The committee does not *at this point* recommend additional direct federal investment in the installation of public charging infrastructure until the relationship between infrastructure availability and PEV adoption and use is assessed. That statement does not mean or should not be construed to mean that no federal investment or additional public infrastructure is needed. Other entities—including vehicle manufacturers, utilities, and other private companies—are actively deploying and planning to deploy public infrastructure and have concluded that additional public infrastructure is needed. However, the committee is recommending research to help determine the relationship between charging infrastructure availability and PEV adoption and use. Although some data have been collected through various projects, the data-collection efforts were not designed to understand that fundamental relationship, and the committee cautions against extrapolating findings on the first adopters to the mainstream market. Given the strain on federal resources, the suggested research should help to ensure that limited federal funds are spent so that they will have the greatest impact.

1

Introduction

Plug-in electric vehicles (PEVs) that derive all or some of their propulsion from an external electricity source have received critical attention in recent years. They are especially attractive because they have the potential to reduce greenhouse gas (GHG) emissions and to decrease petroleum consumption substantially, given that light-duty vehicles account for nearly half of the petroleum consumption in the United States today and that electricity is not typically generated from petroleum (EIA 2014). Globally, the demand for PEVs is growing, and some countries see them as an important element of their long-term strategy to meet environmental, economic, and energy-security goals. Although they hold great promise, there are also many barriers to their penetration into the mainstream market. Some are technical, such as the capabilities of current battery technologies that restrict their electric driving range and increase their purchase price compared with conventional vehicles; others are related to consumer behavior and attitudes; and still others are related to developing an infrastructure to support charging of the vehicles and addressing possible effects of the new charging infrastructure on the electric grid. Given the growing concerns surrounding the perceived barriers, Congress in its 2012 appropriations for the Department of Energy (DOE) requested that DOE commission a study by the National Academies to identify market barriers that are slowing the purchase of PEVs and hindering the deployment of supporting infrastructure.[1] Accordingly, the National Research Council (NRC), which is a part of the National Academies, appointed the Committee on Overcoming Barriers to Electric-Vehicle Deployment, which prepared this final report.

HISTORICAL AND POLICY CONTEXT

The PEV is not a new invention of the twenty-first century. In 1900, 28 percent of the passenger vehicles sold in the United States were electric, and about one-third of the vehicles on the road in New York City, Boston, and Chicago were electric (Schiffer et al. 1994). The demise of PEVs resulted from the mass production of an inexpensive gasoline-powered vehicle (the Model T), the invention of an electric starter for the gasoline vehicle (which eliminated the need for a hand-crank), a supply of affordable gasoline, and the development of the national highway system, which allowed long-distance travel (Schiffer et al. 1994). In the 1970s, interest in PEVs resurfaced with the Arab oil embargo and the emerging environmental and energy security concerns. Over the next few decades, interest in PEVs waxed and waned as gasoline prices remained roughly constant. In the 1990s, interest in PEVs was revived by California's zero-emission-vehicle (ZEV) policies but lagged again primarily because battery technology was not as advanced as it is today. Recent advances in battery and other technologies, new federal standards for carbon-dioxide (CO_2) emissions and fuel economy, state requirements for zero-emission vehicles, and the current administration's goal of putting millions of alternative-fuel vehicles on the road have reignited interest in PEVs.

Recent incentives to increase the number of PEVs on the road began with the Emergency Economic Stabilization Act of 2008, which provided a $2,500 to $7,500 tax credit for the purchase of PEVs (Public Law 110-343 §205). The American Recovery and Reinvestment Act of 2009 (Public Law 111-5 §1141) increased incentives for PEVs by expanding the list of vehicles that are eligible for a tax credit. It also appropriated $2 billion in grants for development of electric-vehicle batteries and related components (DOE 2009) and $2.4 billion in loans for electric-vehicle manufacturing facilities (DOE 2011). Along with private investors, DOE has invested $400 million to support infrastructure development, including demonstration projects involving 13,000 PEVs and 22,000 public and private charging points in 20 U.S. cities (DOE 2011). Furthermore, the DOE Office of Energy Efficiency and Renewable Energy (DOE 2013a) and several national laboratories, including Argonne National Laboratory (ANL 2011, 2012, 2013) and the National Renewable Energy Laboratory (NREL 2013), are conducting substantial research and development on electric-drive technologies for PEVs (NRC 2013a).

Various state-level efforts—such as consumer incentives that include tax credits for vehicle purchase, access to carpool lanes, free public parking, and emission-inspection exemptions—are aimed at increasing the number of PEVs

[1] See Consolidated Appropriations Act, 2012, P.L. 112-74, H. Rept. 112-331 (H.Rept. 112-118).

on the road (DOE 2013b). Other efforts, such as reimbursements and tax incentives for purchasing or leasing charging equipment and low-cost loans for installation projects, are aimed at building the charging infrastructure (DOE 2013b). California's ZEV program is particularly important because of the size of the California motor-vehicle market. Each motor-vehicle manufacturer in the state is required to sell at least a minimum percentage of ZEVs (vehicles that produce no exhaust emissions of any criteria pollutant) and transitional ZEVs (vehicles that can travel some minimum distance solely on a ZEV fuel, such as electricity) (13 CCR §1962.1 [2013]). Nine states—Connecticut, Maine, Maryland, Massachusetts, New Jersey, New York, Rhode Island, Vermont, and Oregon—have also adopted the California ZEV program as part of their plans to meet federal ambient air quality standards.

The policies that promote early PEV deployment are aimed at benefits beyond near-term reductions in petroleum consumption and pollutant emissions. The strategy is to speed the long-term process of converting the motor-vehicle fleet to alternative energy sources by exposing consumers now to PEVs, by encouraging governments and service providers to plan for infrastructure, and by encouraging the motor-vehicle industry to experiment with product design and marketing. Gaining a major market share for PEVs will likely require advances in technology to reduce cost and improve performance, but the premise of the early deployment efforts is that market development and technology development that proceed in parallel will lead to earlier mass adoption than if technology advances are required before beginning market development. The early deployment efforts also might speed technology breakthroughs by maintaining visibility and interest in PEVs. The risk entailed by this strategy is that if PEV promotion efforts are premature relative to the development of the technology, the costs of the promotion will have had little benefit in the form of market development.

The motivation for pursuing PEV-deployment policies beyond their near-term benefit can be understood from the findings of another NRC report, *Transitions to Alternative Vehicles and Fuels*. The committee that prepared that report was asked to assess a range of vehicle technology options and to suggest strategies for attaining petroleum consumption and GHG reduction targets of 50 to 80 percent by the 2030-2050 timeframe (NRC 2013b). An important finding of that report is that major policy initiatives—such as tax incentives, subsidies, or regulations—are required to obtain such large-scale reductions. That conclusion is relevant for the current study because it provides context as to why federal policy (or an NRC study) might focus on barriers. If policy makers decide that such major reductions in petroleum consumption or GHG emissions are required to meet environmental and other goals, an understanding of the barriers and the strategies that are needed to overcome them will be required.

THE PLUG-IN ELECTRIC VEHICLE AND CURRENT SALES

This report focuses on light-duty vehicles (passenger cars and light-duty trucks) in the United States and restricts its discussion to PEVs, which include battery electric vehicles (BEVs)[2] and plug-in hybrid electric vehicles (PHEVs).[3] The common feature of these vehicles is that they can charge their batteries by plugging into the electric grid. The distinction between them is that BEVs operate solely on electricity stored in the battery (there is no other power source), and PHEVs have an internal-combustion engine (ICE) that can supplement the electric power train,[4,5] PEVs are often defined by the number of electric miles that they can drive. A BEV that can drive 100 miles on one battery charge is designated as a BEV100; likewise, a PHEV that can drive 40 miles on one battery charge is designated as a PHEV40. A more detailed discussion of PEV technology is provided in Chapter 2 of this report.

Although a few makes and models of PEVs were available in the mid-1990s (for example, the General Motors EV1 and the Honda EV+, released in 1997; see UCS 2014), many consider the December 2010 introduction of the Nissan Leaf and Chevrolet Volt—the first mass-produced PEVs—to be the start of the viable commercial market for PEVs. Every few months, new PEVs have been added to the U.S. market, including a long-range BEV (the Tesla Model S); limited-range BEVs (such as the Daimler Smart EV and the BMW i3); range-extended PHEVs (such as the Ford Fusion Energi and the Ford C-Max Energi); and minimal PHEVs (such as the Toyota Plug-In Prius).[6] Several manufacturers are also selling limited-volume BEVs, including the Ford Focus EV, the Honda Fit EV, the Fiat 500e, and the Chevrolet Spark to meet fuel-efficiency and ZEV regulatory requirements. In addition, a number of PEVs are not yet available in the United States, notably the Mitsubishi Outlander PHEV and a number of Renault BEVs and Volkswagen PHEVs.

Figures 1-1 and 1-2 show monthly sales for BEVs and PHEVs, respectively. PEV sales in the United States were about 56,000 units in 2012, 96,000 units in 2013, and

---

[2] The term *all-electric vehicle* (AEV) is sometimes used instead of BEV.

[3] BEVs and PHEVs need to be distinguished from conventional hybrid electric vehicles (HEVs), such as the Toyota Prius, which was introduced in the late 1990s. HEVs do not plug into the electric grid but power their batteries from regenerative braking and an internal-combustion engine. They are not included in the PEV category and are not considered further in this report except to make a comparison on some issue.

[4] Several design architectures are available for PHEVs, and, depending on the design, the engine may be used to drive the vehicle directly or act as a generator to recharge the battery or both.

[5] PHEVs can use engines powered by various fuels. This report, however, focuses on PHEV engines that are powered by gasoline because they are the ones currently available in the U.S. market.

[6] PEV designations are discussed in detail in Chapter 2 of this report.

120,000 units in 2014 (Inside EVs 2015). Total U.S. vehicle sales in 2014 were nearly 16.5 million, a record year in which people were replacing their vehicles after not buying during the recession (Woodall and Klayman 2015).

In the U.S. market, PEV sales increased from 0.62 percent in 2013 to 0.76 percent in 2014 (Cobb 2014, 2015); total accumulated sales in the United States were about 291,000 vehicles by the close of 2014 (Inside EVs 2015). To put the U.S. sales data in perspective, Figure 1-3 shows that North America accounted for almost half of the world PEV sales in 2013. Worldwide sales of PEVs were about 132,000 in 2012, 213,000 in 2013, and 318,000 in 2014 (Pontes 2015). PEV sales have not yet been reported for some countries so this number could increase slightly.

The rate of market growth over the past 3 years has almost doubled each year, but sales started at a very low level. By way of comparison, hybrid electric vehicles (HEVs) were introduced in 1997 in Japan and in 1999 in the United States. Although HEVs have been more successful in Japan than in the United States—now at 20 percent of the total Japanese light-duty vehicle market (Nikkei Asian Review 2012) and over 50 percent of Toyota's Japanese vehicle sales (Toyota 2014)—it took 13 years for HEVs to exceed 3 percent of annual new light-duty vehicle sales in the United States (Cobb 2013). However, in certain markets, such as California and Washington, HEVs comprise 10 percent of new passenger vehicle sales (see Chapter 3 for a discussion of factors that affect vehicle preferences). Figure 1-4 compares HEV and PEV sales over their first 34 months of having been introduced to the U.S. market and indicates that PEVs are penetrating the market faster than HEVs.

The California market has been particularly important and accounts for over one-third of annual PEV sales. At the close of 2014, PEV sales in California were 3.2 percent of new light-duty vehicle sales and 5.2 percent of new passenger vehicles (CNCDA 2015). California has a long history of strong sales for new vehicle technologies, especially HEVs as noted above. California is a favorable market for PEVs because it has many wealthy buyers of new technology, broad social support for PEVs in light of its history of air pollution, an active regulatory regime with purchase incentives and mandates for reducing carbon emissions and increasing PEV sales, and favorable weather that is easy on battery life and on charge available for vehicle miles. Furthermore, California has had a consistent, long-standing effort to provide basic Web-based and printed information resources on low- and zero-emission vehicles and to hold some ride-and-drive events. Those activities have likely contributed to greater public awareness of PEVs.

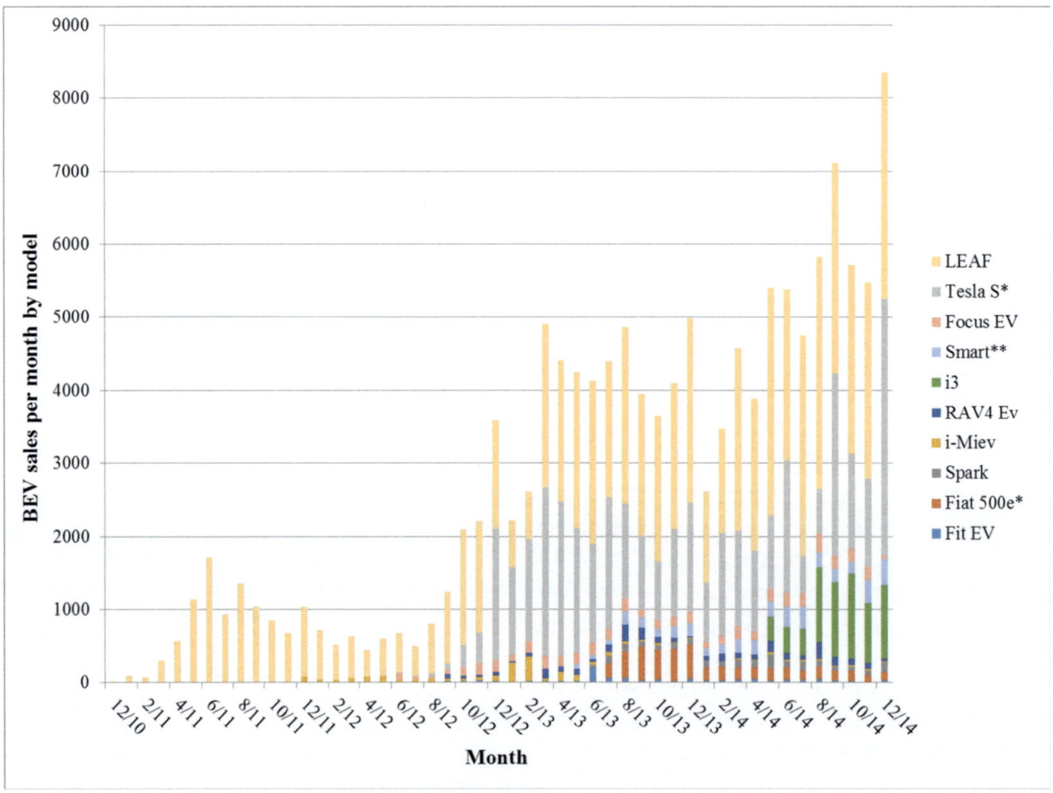

FIGURE 1-1 U.S. BEV monthly sales data from 2010 to 2014. NOTE: BEV, battery electric vehicle. SOURCE: Based on data from Inside EVs (2015).

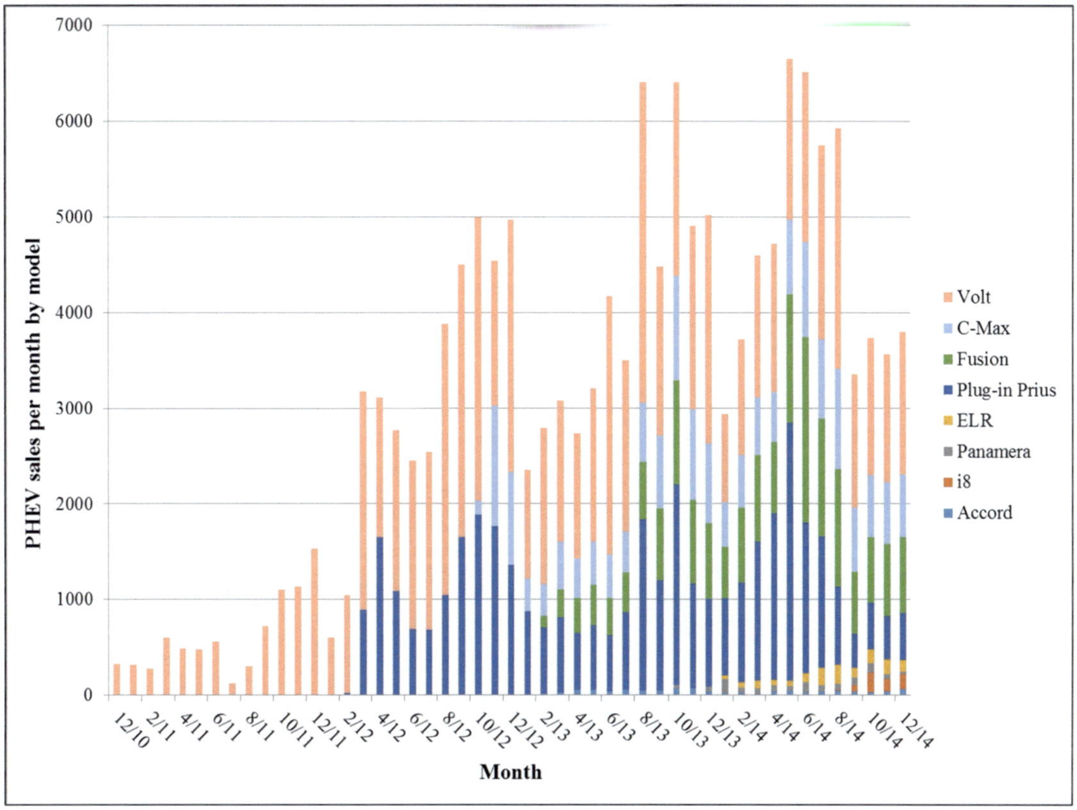

FIGURE 1-2 U.S. PHEV monthly sales data from 2010 to 2014. NOTE: PHEV, plug-in hybrid electric vehicle. SOURCE: Based on data from Inside EVs (2015).

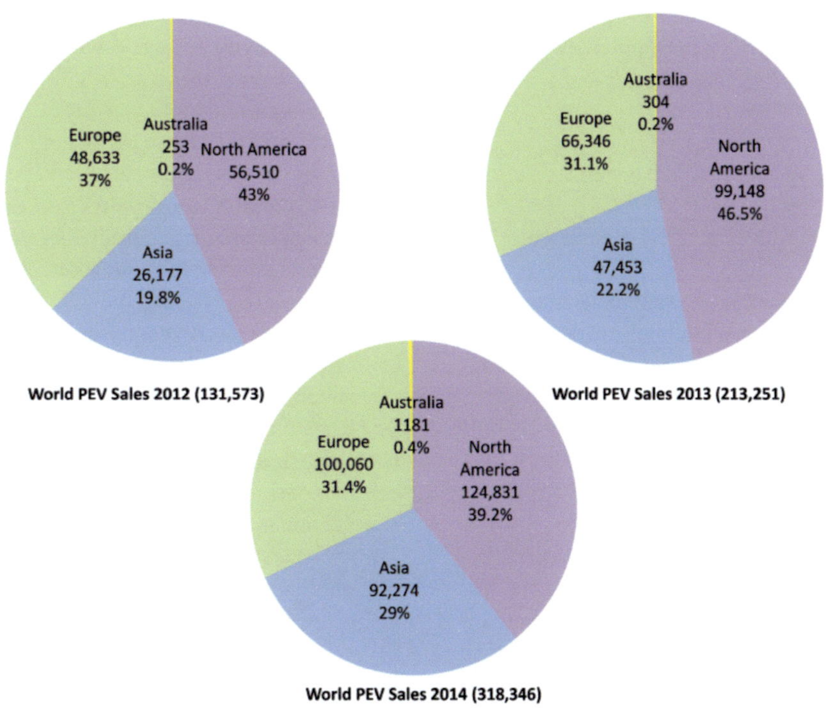

FIGURE 1-3 World PEV sales in 2012, 2013, and 2014. NOTE: PEV, plug-in electric vehicle. SOURCE: Based on data from Pontes (2015).

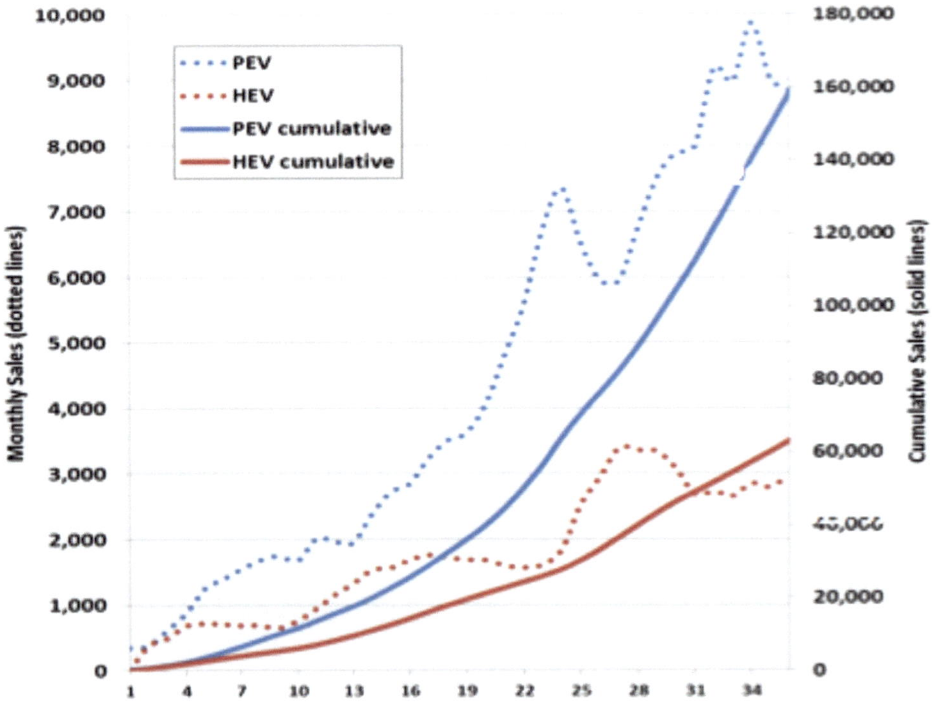

FIGURE 1-4 The rate of PEV market growth in its first 34 months superimposed on the rate of HEV market growth during its first 34 months. NOTE: HEV, hybrid electric vehicle; PEV, plug-in electric vehicle. SOURCE: DOE (2014).

As shown in Figure 1-5, other strong PEV markets are Washington, Oregon, Georgia, Maryland, Vermont, and Hawaii. Those markets have also been driven primarily by social sentiment (an environmentally friendly population base), financial incentives, and regulatory mandates for reducing carbon emissions.

Finding: HEV adoption, which entailed fewer technology changes than PEVs, required 13 years to exceed 3 percent of annual new light-duty vehicle sales in the United States.

Finding: PEVs have had higher sales than HEVs within the first 34 months of their introduction into the market, although the higher sales for PEVs could be the result of the various incentives that have been offered.

PLUG-IN ELECTRIC VEHICLES: BENEFITS AND TRADE-OFFS

PEVs offer several benefits over conventional vehicles. The most obvious for the owner are lower operating cost, less interior noise and vibration from the power train, often better low-speed acceleration, convenient fueling at home, and zero tailpipe emissions when the vehicle operates solely on its battery. BEVs have no conventional transmissions or fuel-injection systems to maintain, do not require oil changes, and have regenerative braking systems that greatly prolong the life of conventional brakes and thus reduce brake repair and replacement costs. On a large scale, PEVs offer the potential to reduce petroleum consumption and improve urban air quality; the degree to which PEVs affect pollutant emissions will depend on how the electricity that fuels a vehicle is generated, the degree to which charging of the vehicle is managed, and the degree to which emissions from power-generation sources are controlled (Peterson et al. 2011; see further discussion below). PEVs might also act as an enabler for renewable power generation by providing storage or rapid demand response through smart-grid applications.

PEVs, however, also have important trade-offs. Current limitations in battery technology result in restricted electric-driving range, high battery cost, long battery-charging time, and uncertain battery life. Concerns about battery safety, depending on the chemistry and energy density of the battery, have also arisen. PEVs have higher upfront costs than their conventional-vehicle counterparts and are available in only a few vehicle models. There is also a need to install a charging infrastructure to support PEVs whether at home, at work, or in a public space. Beyond the technical and economic barriers, people are not typically familiar with the capabilities of PEVs, are uncertain about their costs and benefits, and have diverse needs that current PEVs might not meet. If the goal is widespread deployment of PEVs, it is critical to identify and evaluate the barriers to their adoption.

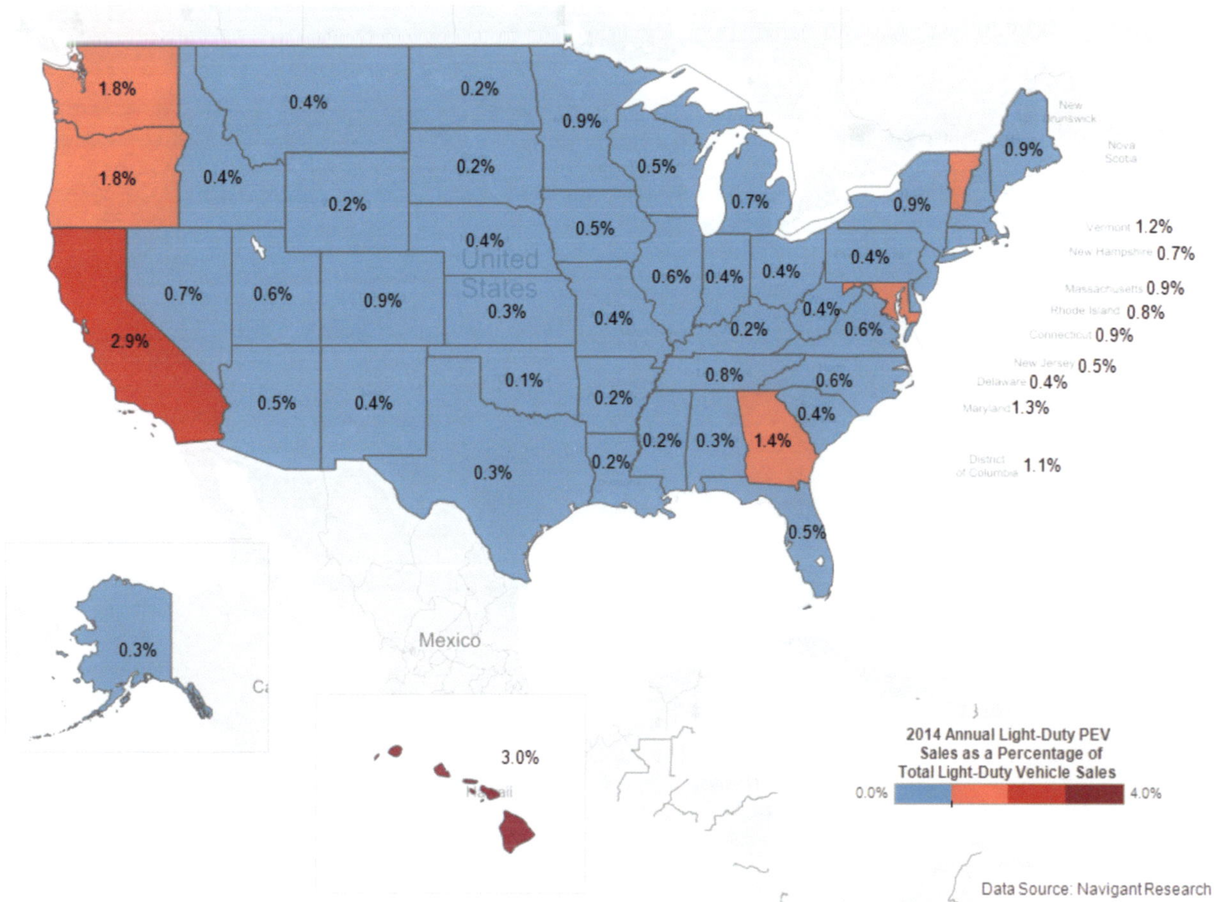

FIGURE 1-5 Projected annual light-duty PEV sales as a percentage of total light-duty vehicle sales. NOTE: PEV, plug-in electric vehicle. SOURCE: Data courtesy of Navigant Research in Shepard and Gartner (2014).

THE COMMITTEE AND ITS TASK

The committee included experts on vehicle technology, electric utilities, business and financial models, economics, public policy, and consumer behavior and response (see Appendix A for biographical information). As noted above, the committee was asked to identify market barriers that are slowing the purchase of PEVs and hindering the deployment of supporting infrastructure in the United States and to recommend ways to mitigate the barriers. The committee's analysis was to be documented in two reports: an interim report and a final comprehensive report. The committee's interim report was released May 2013 and identified infrastructure needs for electric vehicles, barriers to deploying that infrastructure, and possible roles for the federal government in overcoming the barriers. It did not make any recommendations because the committee was in its initial stages of gathering data. After release of the interim report, the committee continued to gather and review information and to conduct analyses. This final comprehensive report addresses the committee's full statement of task, as shown in Box 1-1, and provides recommendations on ways to mitigate the barriers identified.

The premise of the statement of task is that there is a benefit to the United States if a higher fraction of miles driven in the United States is fueled by electricity rather than by petroleum and that PEV deployment will lead to this desired outcome. Two reasons are commonly assumed for the benefit. First, a higher fraction of miles fueled by electricity would reduce U.S. dependence on petroleum. Second, a higher fraction of miles fueled by electricity would reduce the amount of CO_2 and other air pollutants emitted into the atmosphere. The committee was not asked to research and evaluate the premise for the statement of task, and it has not tried to do so. However, it is appropriate to summarize the scientific case on which the premise is based and ask if any recent developments might call the premise into question.

U.S. energy independence and security have been long-term U.S. goals. Every administration from Richard Nixon's onward has proclaimed its importance. A PEV uses no petroleum onboard when it is being fueled by electricity, and in

BOX 1-1 Statement of Task

An ad hoc committee will conduct a study identifying the market barriers slowing the purchase of electric vehicles (EVs, which for this study include pure battery electric vehicles [BEVs] and plug-in hybrid electric vehicles [PHEVs]) and hindering the deployment of supporting infrastructure in the United States. The study will draw on input from state utility commissions, electric utilities, automotive manufacturers and suppliers, local and state governments, the Federal Energy Regulatory Commission, federal agencies, and others, including previous studies performed for the Department of Energy (DOE), to help identify barriers to the introduction of electric vehicles, particularly the barriers to the deployment of the necessary vehicle charging infrastructure, and recommend ways to mitigate these barriers. The study will focus on light-duty vehicles but also draw upon experiences with EVs in the medium- and heavy-duty vehicle market segment. Specifically, the committee will:

1. Examine the characteristics and capabilities of BEV and PHEV technologies, such as cost, performance, range, safety, and durability, and assess how these factors might create barriers to widespread deployment of EVs. Included in the examination of EV technologies will be the characteristics and capabilities of vehicle charging technologies.
2. Assess consumer behaviors and attitudes towards EVs and how these might affect the introduction and use of EVs. This assessment would include analysis of the possible manner by which consumers might recharge their vehicles (vehicle charging behaviors, e.g., at home, work, overnight, frequency of charging, time of day pricing, during peak demand times, etc.) and how consumer perceptions of EV characteristics will impact their deployment and use.
3. Review alternative scenarios and options for deployment of the electric vehicle infrastructure, including the various policies, including tax incentives, and business models necessary for deploying and maintaining this infrastructure and necessary funding mechanisms. The review should include an evaluation of the successes, failures, and lessons learned from EV deployment occurring both within and outside the United States.
4. Examine the results of prior (and current) incentive programs, both financial and other, to promote other initially uneconomic technologies, such as flex-fuel vehicles, hybrid electric vehicles, and now PHEVs/BEVs to derive any lessons learned.
5. Identify the infrastructure needs for the electricity sector, particularly the needs for an extensive electricity charging network, the approximate costs of such an infrastructure, and how utility investment decision making will play into the establishment of a charging network. As part of this assessment, the committee will identify the improvements in the electricity distribution systems needed to manage vehicle charging, minimize current variability, and maintain power quality in the local distribution network. Also, the committee will consider the potential impacts on the electricity system as a whole, potentially including: impacts on the transmission system; dispatch of electricity generation plants; improvements in system operation and load forecasting; and use of EVs as grid-integrated electricity storage devices.
6. Identify the infrastructure needs beyond those related to the electricity sector. This includes the needs related to dealer service departments, independent repair and maintenance shops, battery recycling networks, and emergency responders.
7. Discuss how different infrastructure deployment strategies and scenarios might impact the costs and barriers. This might include looking at the impacts of focusing the infrastructure deployment on meeting the needs for EVs in vehicle fleets, where the centralization of the vehicle servicing might reduce the costs for deploying charging infrastructure or reduce maintenance issues, or focusing the infrastructure deployment on meeting the needs for EVs in multi-family buildings and other high-density locations, where daily driving patterns may be better suited to EV use than longer commutes from single family homes in lower density areas. This might also include looking to the extent possible of how the barriers and strategies for overcoming barriers may differ in different U.S. localities, states, or regions.
8. Identify whether there are other barriers to the widespread adoption of EVs, including shortages of critical materials, and provide guidance on the ranking of all barriers to EV deployment to help prioritize efforts to overcome such barriers.
9. Recommend what roles (if any) should be played by the federal government to mitigate those market barriers and consider what federal agencies, including the DOE, would be most effective in those roles.
10. Identify how the DOE can best utilize the data on electric vehicle usage already being collected by the department.

The committee's analysis and methodologies will be documented in two NRC-approved reports. The study will consider the technological, infrastructure, and behavioral aspects of introducing more electric vehicles into the transportation system. A short interim report will address, based on presentations to the committee and the existing literature, the following issues:

1. The infrastructure needs for electric vehicles;
2. The barriers to deploying that infrastructure; and
3. Optional roles for the federal government to overcome these barriers, along with initial discussion of the pros and cons of these options.

The final report will discuss and analyze these issues in more detail and present recommendations on the full range of tasks listed in Items (1) to (10) for the full study. The final report will include consideration of the infrastructure requirements and barriers as well as technological, behavioral, economic, and any other barriers that may slow the deployment of electric vehicles, as well as recommendations for mitigating the identified market barriers. It is envisioned that the committee will hold meetings in different locations around the United States, as well as collect information on experiences in other countries, in order to collect information on different approaches being taken to overcoming the barriers to electric vehicle deployment and its supporting charging infrastructure.

2013, less than 0.7 percent of the U.S. grid electricity was produced from petroleum.[7] Thus, widespread adoption of PEVs would lead to a large decrease in petroleum use. There is a modest caveat, however, to that conclusion. U.S. petroleum consumption in the light-duty vehicle fleet is regulated by National Highway Traffic Safety Administration (NHTSA) through its Corporate Average Fuel Economy (CAFE) program (see Chapter 7 for a detailed discussion). CAFE standards are based on average fuel economy of a manufacturer's vehicle fleet, so reductions in fuel use attributed to the sale of a single PEV could be offset by the sale of an ICE vehicle[8] that consumes more fuel, resulting in no net fuel savings from PEV deployment (Gecan et al. 2012). However, petroleum consumption might still be reduced by PEV deployment because the CAFE program underestimates the petroleum-reduction benefit of PEVs. Specifically, the factor used by the CAFE program to calculate a fuel-economy rating for compliance is equivalent to assuming that 15 percent of the electrical energy used by a PEV is generated from petroleum, which is clearly an overestimate of the petroleum used by the U.S. electric sector (EPA/NHTSA 2012, p. 62821). Moreover, successful deployment of PEVs would help to enable the implementation of increasingly stringent CAFE standards, resulting in lower petroleum consumption, as noted by the Congressional Budget Office (Gecan et al. 2012).

In addition to reduced petroleum consumption, lower GHG emissions are noted as a reason for PEV deployment. A series of authoritative scientific reports (IPCC 2014; NCA 2014; NRC 2014) stress that the emission of GHGs, particularly CO_2, is contributing in a measurable way to global warming and urge the United States to reduce its CO_2 emissions. Because light-duty vehicles were responsible for 17.4 percent of total U.S. GHG emission in 2012 (EPA 2014a), reducing GHG emissions from the light-duty vehicle fleet is seen as an important approach for reducing overall GHG emissions. A vehicle completely powered by electricity from the U.S. electric grid is often called a zero-emission vehicle (ZEV) insofar as it emits no CO_2 or other pollutants from its tailpipe. However, whether PEVs reduce total U.S. emissions of CO_2 and other GHGs depends on the emissions associated with the production of the grid electricity that the vehicles use and, in the case of PHEVs, on tailpipe emissions. Estimation of the emissions attributed to a vehicle whether operating on gasoline or electricity is often referred to as a well-to-wheels analysis.[9] For a gasoline vehicle, a well-to-wheels analysis would consider emissions from fossil fuel extraction, refining, and transportation, as well as tailpipe emissions from onboard fuel combustion. For a PEV, a well-to-wheels analysis would include emissions associated with electricity generation, such as extraction of fuels, their transportation, and the transmission of the electricity. For PHEVs, a well-to-wheels analysis would be a weighted average of the emissions from electricity-fueled and petroleum-fueled operation.

There are several (often conflicting) methods to evaluate well-to-wheels GHG emissions of vehicles. One method is to use well-to-wheels emission factors produced by DOE. Given that method, an analysis of the 30 mpg 2014 Chevrolet Cruze (an ICE vehicle), the 50 mpg 2014 Toyota Prius (one of the cleanest HEVs), and the Nissan Leaf BEV charged on the 2010 U.S. average electricity-generation mix shows that the Cruze, Prius, and Leaf produce GHGs of 369 g/mi, 222 g/mi, and 200 g/mi, respectively.[10] Accordingly, the operation of the BEV is estimated to produce about 46 percent less GHG than the ICE vehicle and 10 percent less GHG than the best hybrid. If one considered cleaner electricity sources (for example, ones in California or Washington, where large numbers of PEVs are purchased), the BEV would produce only about half of the GHG of the best HEV (DOE 2015). Well-to-wheels analyses of this type have been reported for average GHG emissions within each grid subregion as defined by the U.S. Environmental Protection Agency (EPA) (Anair and Mahmassani 2012).

An alternative analysis examines the emissions attributed to PEV charging by taking into account not only the average emissions at a given location, but also the variation in emissions due to time of day and the type of generation added to provide the additional electricity needed for charging. Analyses of this type differ on the emissions resulting from PEVs, depending on the modeling approach and the time frame used. On the one hand, EPA in its latest rulemaking for light-duty CO_2 standards found that the additional power plants used to meet PEV load in the 2022-2030 time frame would have lower emissions than the national average power plant at that time (EPA/NHTSA 2012, p. 62821). On the other hand, a model that attempts to simulate emissions from today's grid using older data from 2007 to 2009 suggests that the marginal emission rates for PEV charging might be higher than the average power plant emissions and in the worst case might even be higher than emissions attributed to HEVs and ICE vehicles (Graff Zivin 2014).

Another factor to consider is the treatment of GHG emissions from PEVs under the joint CAFE-GHG standards (see Chapter 7 for a more detailed discussion). Similar to the CAFE program requirement for a fleetwide average fuel economy, fleetwide average GHG emission rates are restricted to a certain average grams of CO_2 per mile. Therefore, lower PEV emission rates are averaged with higher emission rates from ICE vehicles. If, however, standards become increasingly more stringent, PEV sales might be needed

[7] Estimate calculated from data reported in EIA (2013), Short Term Energy Outlook.

[8] For this report, *ICE vehicle* or *conventional vehicle* refers to a light-duty vehicle that obtains all of its propulsion from an internal-combustion engine.

[9] A more complete analysis is a lifecycle assessment that, in addition to the well-to-wheels assessment, includes environmental impacts from vehicle production (all aspects), vehicle use, and disposal of the vehicle at the end of its life.

[10] The latest data for ICE tailpipe emissions and for the "upstream emissions" of GHGs (CO_2 equivalent) to produce electricity from the 2010 U.S. electricity grid are available at www.fueleconomy.gov.

to meet them, and early deployment of PEVs encouraged through incentives might allow the implementation of more stringent GHG standards in the future. To encourage PEV deployment in the near term, EPA temporarily allows the portion of PEV miles that are estimated to be driven on electricity to be treated as zero emissions and lets a single PEV count as more than one vehicle. That favorable treatment creates a short-term trade-off in GHG emissions that is anticipated to bring long-term benefits from PEV deployment.

Emissions attributed to PEV operation might change over time with changes in emissions from electricity generation. The United States has reduced its GHG emissions over the last several years by converting some of its electricity production from coal to natural gas. The result is that, on average, a PEV fueled by electricity is now responsible for less GHG per mile driven. Well-to-wheels emissions must continue to consider the evolving understanding of upstream methane emissions from coal and natural gas production and distribution (EPA 2014b). The substantial reductions in U.S. GHG emissions from electricity generation are expected to continue for some time, especially if the proposed EPA GHG regulations of new and existing power plants and oil and gas wells are enacted. Thus, PEVs will make further reductions in GHG emissions as the U.S. electric grid changes to lower carbon sources for its electricity—a fact that is sometimes overlooked. And PEVs fueled on electricity have the potential to produce no well-to-wheels emissions if the electricity is generated from carbon-free sources. That is not the case for even the most efficient petroleum-fueled ICE vehicles. If the United States intends to reach low levels of GHG emissions (80 percent reduction), large-scale adoption of PEVs is one viable option (NRC 2013b).

The committee concludes that the premise for the statement of task—that there is an advantage to the United States if a higher fraction of the miles driven here is fueled from the U.S. electric grid—is valid now. The advantage becomes even greater each year that the United States continues to reduce the GHGs that it produces in generating electricity.

Finding: The average GHG emissions for which PEVs are responsible are currently lower than emissions from even the cleanest gasoline vehicles and will be further reduced as the electricity for the U.S. grid is produced from lower carbon sources.

Recommendation: As the United States encourages the adoption of PEVs, it should continue to pursue in parallel the production of U.S. electricity from increasingly lower carbon sources.

The committee notes that the use of HEVs rather than ICE vehicles would provide a large reduction in U.S. petroleum use and emissions. If their small market share could be substantially increased, the many types of HEVs already on the market could rapidly bring about substantial reductions in petroleum use and emissions in the time that a comparable variety of PEVs are brought to market. Accordingly, the focus in this report on PEVs should not be misinterpreted so as to keep policy makers from encouraging the switch from ICE vehicles to HEVs.

THE COMMITTEE'S APPROACH TO ITS TASK

Ten meetings were held over the course of this study. Seven meetings included open sessions during which the committee heard from the sponsor and invited speakers representing national laboratories, state agencies, university centers, vehicle manufacturers and dealers, and other private industries and consultants (see Appendix B for a list of speakers from all the open sessions). Committee subgroups also visited several sites in this country and abroad, including Texas, Japan, Germany, and the Netherlands, to gather information on electric-vehicle programs. On those trips, the committee members met with national and regional government officials, automobile manufacturers, charging companies, and other relevant organizations. On the basis of information received at its meetings, its on-site visits, and from the literature, the committee prepared this final report.

As discussed above, the committee accepted its charge and is not debating the merits of promoting, enabling, or increasing PEV adoption. This report focuses on ways to extend the market from "early adopters" to more mainstream customers. Early-market customers for PEVs tend to base their purchase decisions more on personal values and less on purchase price. In contrast, mainstream-market customers tend to weigh price and overall vehicle utility more heavily in their purchase decisions.

One final issue concerns the rapidly changing market and the various factors that hinder the adoption of PEVs—particularly the price of gasoline. Wide fluctuations in gasoline prices, as occurred over the course of this study, affected the committee's comparisons and conclusions about the cumulative costs of vehicle ownership. As discussed in Chapter 7, gasoline prices are an important factor in determining the benefits of PEV ownership and can provide an incentive or a disincentive for purchasing a PEV. To address the issue of fluctuating gasoline prices, the committee decided that the best approach was to use a range of gasoline prices, from \$2.50 to \$4.00, in its calculations, to present ranges as appropriate throughout its report, and to draw conclusions based on these ranges.

ORGANIZATION OF THIS REPORT

This final report is organized into seven chapters and three appendixes. Chapter 2 discusses the current characteristics and capabilities of PEV technologies. Chapter 3 provides a brief assessment of consumer behavior and attitudes toward PEVs and how they are affecting PEV deployment. Chapter 4 discusses what can be done to improve institutional support for PEV deployment. Chapter 5 provides an in-depth discussion of the charging infrastructure needed for PEV deployment, and Chapter 6 evaluates the ability of the electric infrastructure to meet the increased electricity demand in light of the new charging infrastructure. Chapter 7 discusses ways to motivate the consumer. Appendix A provides biographical information for committee members, Appendix B lists the meetings and the presentations made in open sessions, and Appendix C provides some information on international programs to support PEV deployment.

REFERENCES

Anair, D., and A. Mahmassani. 2012. State of Charge: Electric Vehicles' Global Warming Emissions and Fuel-Cost Savings across the United States. Union of Concerned Scientists. http://www.ucsusa.org/assets/documents/clean_vehicles/electric-car-global-warming-emissions-report.pdf.

ANL (Argonne National Laboratory). 2011. "Hybrid Vehicle Technology." http://www.transportation.anl.gov/hev/index.html. Accessed March 14, 2013.

ANL. 2012. "Advanced Battery Research, Development, and Testing." http://www.transportation.anl.gov/batteries/index.html. Accessed March 14, 2013.

ANL. 2013. "Argonne Leads DOE's Effort to Evaluate Plug-in Hybrid Technology." http://www.transportation.anl.gov/phev/index.html. Accessed March 14, 2013.

CNCDA (California New Car Dealers Association). 2015. California Auto Outlook: Fourth Quarter 2014. Volume 11, Number 1, February. http://www.cncda.org/CMS/Pubs/Cal_Covering_4Q_14.pdf.

Cobb, J. 2013. "December 2012 Dashboard." Hybrid Cars, January 8. http://www.hybridcars.com/december-2012-dashboard.

Cobb, J. 2014. "August 2014 Dashboard." Hybrid Cars, September 4. http://www.hybridcars.com/august-2014-dashboard/.

Cobb, J. 2015. "December 2014 Dashboard." Hybrid Cars, January 6. http://www.hybridcars.com/december-2014-dashboard/.

DOE (U.S. Department of Energy). 2009. Recovery Act—Electric Drive Vehicle Battery and Component Manufacturing Initiative. Funding Opportunity No. DE-FOA-0000026. http://www1.eere.energy.gov/vehiclesandfuels/pdfs/de-foa-0000026-000001.pdf.

DOE. 2011. One Million Electric Vehicles by 2015: February 2011 Status Report. http://www1.eere.energy.gov/vehiclesandfuels/pdfs/1_million_electric_vehicles_rpt.pdf.

DOE. 2013a. "Hybrid and Vehicle Systems." Vehicle Technologies Office. http://www1.eere.energy.gov/vehiclesandfuels/technologies/systems/index.html. Accessed March 14, 2013.

DOE. 2013b. "State Laws and Incentives." Alternative Fuels Data Center. http://www.afdc.energy.gov/laws/state. Accessed January 29, 2013.

DOE. 2014. Presentation at the NEXTSteps Symposium, University of California, Davis.

DOE. 2015. "Greenhouse Gas Emissions for Electric and Plug-In Hybrid Electric Vehicles." http://www.fueleconomy.gov/feg/Find.do?zipCode=90001&year=2014&vehicleId=34699&action=bt3.

EIA (U.S. Energy Information Administration). 2013. Short-Term Energy Outlook. http://www.eia.gov/forecasts/steo/archives/dec13.pdf.

EIA. 2014. Annual Energy Outlook 2014. http://www.eia.gov/forecasts/aeo/. Accessed September 24, 2014.

EPA (U.S. Environmental Protection Agency). 2014a. Inventory of U.S. Greenhouse Gas Emissions and Sinks: 1990-2012. EPA 430-R-14-003. http://www.epa.gov/climatechange/Downloads/ghgemissions/US-GHG-Inventory-2014-Main-Text.pdf.

EPA. 2014b. Regulatory Impact Analysis for the Proposed Carbon Pollution Guidelines for Existing Power Plants and Emission Standards for Modified and Reconstructed Power Plants. EPA-452/R-14-002. http://www2.epa.gov/sites/production/files/2014-06/documents/20140602ria-clean-power-plan.pdf.

EPA/NHTSA (U.S. Environmental Protection Agency and National Highway Traffic Safety Administration). 2012. 2017 and Later Model Year Light-Duty Vehicle Greenhouse Gas Emissions and Corporate Average Fuel Economy Standards. Federal Register 77(199): October. http://www.gpo.gov/fdsys/pkg/FR-2012-10-15/pdf/2012-21972.pdf.

Gecan, R., J. Kile, and P. Beider. 2012. Effects of Federal Tax Credits for the Purchase of Electric Vehicles. Congressional Budget Office. http://www.cbo.gov/sites/default/files/cbofiles/attachments/09-20-12-ElectricVehicles_0.pdf.

Graff Zivin, J.S., M.J. Kotchen, and E.T. Mansur. 2014. Spatial and temporal heterogeneity of marginal emissions: implications for electric cars and other electricity-shifting policies. Journal of Economic Behavior & Organization 107: 248-268.

Inside EVs. 2015. "Monthly Plug-in Sales Scorecard." http://insideevs.com/monthly-plug-in-sales-scorecard/.

IPCC (Intergovernmental Panel on Climate Change). 2014. Fifth Assessment Report. http://www.ipcc.ch/report/ar5/index.shtml.

NCA (National Climate Assessment). 2014. "Climate Change Impacts in the United States." U.S. Global Change Research Program, Washington, DC. http://nca2014.globalchange.gov.

Nikkei Asian Review. 2012. "Hybrids 19.7% of New Cars Sold in May 2012 in Japan." Integrity Exports, June 8. http://integrityexports.com/2012/06/08/hybrids-19-7-of-new-cars-sold-in-ma-in-japan/. Accessed June 10, 2012.

NRC (National Research Council). 2013a. Review of the Research Program of the U.S. DRIVE Partnership: Fourth Report. Washington, DC: The National Academies Press.

NRC. 2013b. Transitions to Alternative Vehicles and Fuels. Washington, DC: The National Academies Press.

NRC. 2014. Climate Change: Evidence and Causes. Washington, DC: The National Academies Press.

NREL (National Renewable Energy Laboratory). 2013. "Plug-In Hybrid Electric Vehicles." Vehicle Systems Analysis. http://www.nrel.gov/vehiclesandfuels/vsa/plugin_hybrid.html. Accessed March 14, 2013.

Peterson, S.B., J.F. Whitacre, and J. Apt. 2011. Net air emissions from electric vehicles: The effect of carbon price and charging strategies. Environmental Science and Technology 45(5): 1792-1797.

Pontes, J. 2015. "World Top 20 December 2014." EV Sales, January 31. http://ev-sales.blogspot.com/2015/01/world-top-20-december-2014-special.html.

Schiffer, M.B., T.C. Butts, and K.K. Grimm. 1994. Taking Charge: The Electric Automobile in America. Washington, DC: Smithsonian Institution Press.

Shepard, S., and J. Gartner. 2014. Electric Vehicle Geographic Forecasts. Navigant Research. Boulder, CO.

Toyota. 2014. "Worldwide Sales of Toyota Hybrids Top 6 Million Units." News release. January 14. http://corporatenews.pressroom.toyota.com/releases/worldwide+toyota+hybrid+sales+top+6+million.htm.

UCS (Union of Concerned Scientists). 2014. "Electric Vehicle Timeline: Electric Cars, Plug-In Hybrids, and Fuel Cell Vehicles." http://www.ucsusa.org/clean_vehicles/smart-transportation-solutions/advanced-vehicle-technologies/electric-cars/electric-vehicle-timeline.html.

Woodall, B., and B. Klayman. 2015. U.S. auto sales end 2014 strong but slower growth looms. Reuters, January 5. http://www.reuters.com/article/2015/01/05/us-autos-sales-usa-idUSKBN0KE0WH20150105.

2

Plug-in Electric Vehicles and Charging Technologies

As discussed in Chapter 1, the assigned task for the present report is to examine barriers to the adoption of plug-in electric vehicles (PEVs), which use electricity from the U.S. electric grid as their fuel. When powered by electricity from the grid, which uses little petroleum to produce electricity, such vehicles require essentially no petroleum, and they emit no carbon dioxide (CO_2) or other harmful pollutants from the tailpipe (EPA 2012). The premise for the assigned task is that such vehicles have the potential for significantly lowering petroleum consumption and decreasing emissions now and even more so in the future. The CO_2 emissions advantage will grow as the United States continues to switch to lower-carbon-emitting sources of electricity by phasing out coal and natural gas combustion and replacing those energy sources with solar, wind, or nuclear energy, or alternatively by using carbon capture and sequestration for coal and natural gas plants if that technology ever proves to be practical.

As described in more detail in this chapter, electricity from a battery powers the electric motor of a PEV and is thus the analog of the gasoline in a tank that powers the internal-combustion engine (ICE) of a conventional vehicle. The hundreds of miles of range that is typical for a conventional vehicle depends on how many gallons of fuel the tank can hold and on the fuel economy of the vehicle. Similarly, the all-electric range (AER) of a vehicle (the distance that it can travel fueled only by the electricity that can be stored in its battery) depends on the size of the battery and the efficiency of the vehicle. The AER, like the range of an ICE vehicle, depends on such factors as the vehicle design, including its aerodynamics, rolling resistance, and weight; the driving environment, including road grade and outside temperature; the amount of heating and cooling that is used; and how aggressively the vehicle is driven (NREL 2013). Some factors, such as outside temperature, will have a greater effect on PEVs than ICE vehicles. The ranges quoted in the present report are taken from the U.S. Environmental Protection Agency (EPA) data on results from the standard driving cycle (EPA 2014).

This chapter begins with a discussion of the capabilities and limitations of four classes of PEVs, each presenting different obstacles to widespread consumer adoption. It continues with a discussion of high-energy batteries, the critical and expensive components for all PEVs, and the possibility of increasing the energy densities of these batteries. A summary of current and projected battery costs is provided because it is primarily higher battery costs that make PEVs cost more than ICE vehicles. The chapter concludes with a discussion of vehicle charging and charging options. The committee's findings and recommendations are provided throughout this chapter.

TYPES OF PLUG-IN ELECTRIC VEHICLES

Essentially all U.S. vehicles today have an ICE that uses gasoline or diesel fuel that is derived from petroleum and produces CO_2 and other harmful emissions as the vehicles travel. Hybrid electric vehicles (HEVs) achieve a lower fuel consumption than conventional vehicles of the same size and performance. They typically have a smaller ICE and a high-power battery and electric motor to increase the vehicle's acceleration when needed and to power the vehicle briefly at low speeds. Electric energy is provided to the battery when the vehicle brakes and is produced by the ICE using power that is not needed to propel the vehicle. The lower fuel consumption that can be achieved is illustrated by the 50 miles per gallon (mpg) of gasoline that is achieved by the Toyota Prius, the best-selling HEV. There are many other HEV models available in the market, most of which use much less fuel than their ICE counterparts. Although HEVs still constitute a small fraction of the U.S. vehicle fleet, the more rapid adoption of efficient HEVs could be important for meeting the increasingly stringent corporate average fuel economy (CAFE) and greenhouse gas (GHG) emission standards that are helping to drive down the demand for petroleum and to decrease vehicle tailpipe emissions. However, although HEVs use batteries and electric motors, they derive all of their electric and mechanical energy from their gasoline or diesel fuel. Thus, HEVs are used as a point of comparison for the present report, but they are not its primary focus.

As noted in Chapter 1, the PEVs that are the focus of the present report are often divided into two categories: battery electric vehicles (BEVs) and plug-in hybrid electric vehicles (PHEVs) that include an ICE and an electric motor. This chapter uses vehicle AER to distinguish four classes of PEVs. The reason is that the obstacles to consumer adoption and the charging infrastructure requirements differ for the four classes of PEVs. BEVs are separated into long-range BEVs and limited-range BEVs, and PHEVs are separated into range-extended PHEVs and minimal PHEVs.

There are now examples in the market for each type of PEV, and the committee uses some of them to illustrate their capabilities (see Table 2-1). Despite the increasing number of PEVs entering the market, however, far fewer vehicle types and features are available compared with the types and features available for conventional ICE vehicles and HEVs. Chapter 3 discusses the current paucity of choices as a possible barrier to PEV adoption. As PEVs become more common, however, the variety of choices will increase, and some models could emerge that do not fit perfectly into one of the four categories described here.

Finding: The increasing number of PEVs entering the market demonstrates the possibility of various types of electrically fueled vehicles, although far fewer vehicle types and

TABLE 2-1 Definitions and Examples of the Four Types of Plug-in Electric Vehicles

Vehicle	Battery Capacity[a]	All-Electric Range[b]
Type 1. Long-Range Battery Electric Vehicle. Can travel hundreds of miles on a single battery charge and then be refueled in a time that is much shorter than the additional driving time that the refueling allows, much like an ICE vehicle or HEV.		
 **2014 Tesla Model S** © Steve Jurvetson, licensed under Creative Commons 2.0 (CC-BY-2.0)	85 kWh nominal	265 miles
Type 2. Limited-Range Battery Electric Vehicle. Is made more affordable than the long-range BEV by reducing the size of the high-energy battery. Its limited range more than suffices for many commuters, but it is impractical for long trips.		
 **2014 Nissan Leaf** ©2014 Nissan North America, Inc. Nissan, Nissan model names, and the Nissan logo are registered trademarks of Nissan	24 kWh nominal (~21 kWh usable)	84 miles
 **2014 Ford Focus Electric** Image courtesy of Ford Motor Company	23 kWh nominal	76 miles
Type 3. Range-Extended Plug-in Hybrid Electric Vehicle. Operates as a zero-emission vehicle until its battery is depleted, whereupon an ICE turns on to extend its range.		
 **2014 Chevrolet Volt** © General Motors	16.5 kWh nominal (~11 kWh usable)	38 miles
Type 4. Minimal Plug-in Hybrid Electric Vehicle. Is mostly an HEV. Its small battery can be charged from the grid, but it has an all-electric range that is much smaller than the average daily U.S. driving distance.		
 **2014 Toyota Plug-in Prius** Image courtesy of Toyota Motor Corporation	4.4 kWh nominal (~3.2 kWh usable)	11 miles (blended) 6 miles (battery only)

[a] Nominal battery capacities, reported by manufacturers in product specifications, are for a battery before it goes into a vehicle. Vehicle electronics restrict the usable battery capacity to what becomes the vehicle's all-electric range.
[b] The all-electric ranges noted are average values estimated by EPA. The motor size and design architecture of the Toyota Plug-in Prius require the use of its ICE to complete the Federal Test Procedure; therefore, its range is given for both blended, charge-depleting operation and battery-only operation. All other vehicle ranges are given only for fully electric, charge-depleting operation. NOTE: HEV, hybrid electric vehicle; ICE, internal-combustion engine.
SOURCES: Based on data from Duoba (2012); DOE/EPA (2014a, 2014b, 2014c, 2014d, 2014e); DOE (2012, 2013); EPA (2014); Ford (2014); and Toyota (2014).

features are currently available than are available for conventional ICE vehicles and HEVs.

Type 1: Long-Range Battery Electric Vehicles

Today's drivers are accustomed to ICE and HEV vehicles that are able to drive for hundreds of miles and then be refueled at any gasoline station in several minutes. Extended trips are practical insofar as the refueling time is much shorter than the additional driving time that refueling provides. The full-size Tesla Model S is a demonstration that hundreds of miles are also possible with a BEV that gets its energy entirely from the electric grid. It has a range based on the EPA driving cycle of 265 miles for a single charge of its 85 kWh battery (DOE/EPA 2014a). Half of the charge of a depleted battery can be replenished in 20 minutes at any of the superchargers that Tesla is installing for its customers along major U.S. highways. That charge would extend the driving distance by about 132 miles. Thus, the Tesla Model S is considered a long-range BEV because it can drive for hundreds of miles on a charge and then be refueled in a time that is much shorter than the additional driving time that the refueling allows. Although filling a vehicle with gasoline or diesel would be much quicker, the ability to travel almost 400 miles stopping only once for a 20-minute recharge is a notable achievement for a BEV. With its high acceleration performance, low noise, high-end styling, and expected low maintenance, the Tesla Model S has earned several consumer performance awards (MacKenzie 2013; Consumer Reports 2014).

The Tesla Model S is priced as a high-end luxury vehicle comparable to a high-end BMW and is not affordable for most U.S. drivers.[1] Nonetheless, it is an important demonstration of the possibility of a long-range BEV for consumers. For now, however, high battery cost is a barrier to the mass adoption of the Tesla Model S and other BEVs. The fuel cost per mile and maintenance costs are much smaller for BEVs than for ICE vehicles, but not enough to offset their higher purchase price at current U.S. petroleum prices. The situation can be quite different in countries where gasoline and diesel fuel cost 2 or 3 times as much as in the United States.

Finding: The possibility of a long-range BEV that is powered by grid electricity rather than gasoline or diesel and that meets consumer performance needs has been clearly demonstrated by the full-size Tesla Model S.

Type 2: Limited-Range Battery Electric Vehicles

The high cost of high-energy batteries leads to three types of more affordable PEVs. The first sacrifices driving range and the other two sacrifice zero tailpipe emissions for longer trips. A limited-range BEV is more affordable simply because a smaller high-energy battery is installed, giving it a shorter range. The 2014 Nissan Leaf, a midsize car, is the best-selling example. It has a 24 kWh battery and an 84-mile range (DOE/EPA 2014b). A more recent addition to the limited-range BEV market is the Ford Focus Electric compact car, which has a 76-mile range (DOE/EPA 2014c). As noted earlier in this chapter, the actual range of a BEV will depend on a variety of factors, including climate, road grade, and driver behavior. The difference between the range, fuel economy, and emission performance estimated for regulatory compliance and what is actually experienced by drivers of all types of light-duty vehicles continues to be controversial and is discussed in other NRC reports (NRC 2011, 2013).

The ranges that are achievable by limited-range BEVs are much longer than the 40 or fewer miles that 68 percent of U.S. drivers drive in a day, making these vehicles adequate for normal commuting and the average daily use (FHWA 2011). However, drivers of ICE vehicles are accustomed to being able to travel well beyond the average daily distance when the need arises and can add hours of additional traveling time by simply refilling a gasoline or diesel fuel tank in several minutes. For a limited-range BEV, however, a half hour of the fastest available charging will typically allow an hour or even less of additional driving, making extended trips impractical. For extended trips and driving distances much beyond the AER, the limited-range BEV driver needs to have access to a second vehicle that has no serious range limitations or to some other transportation means. As discussed in Chapter 3, many households have two or more vehicles, so trading vehicle utility within a household is already common. For its customers, BMW is experimenting with offering access to an ICE vehicle for the occasional long trip to see if this perk lowers the barrier to adoption of its vehicles. Rental companies like Hertz have also indicated that they are interested in filling that same niche (Hidary 2012).

Finding: Limited-range BEVs are the only type of PEV that have a considerable range limitation. However, the range that they do have more than suffices for the average daily travel needs of many U.S. drivers.

Finding: Given the substantial refueling time that would be required, limited-range BEVs are not practical for trips that would require more than one fast charge.

Type 3: Range-Extended Plug-in Hybrid Electric Vehicles

A range-extended PHEV[2] is similar to a long-range or limited-range BEV in that the battery can be charged from

[1] The cost of producing a Model S is currently offset somewhat in that Tesla is able to sell the zero-emission-vehicle (ZEV) credit it earns for each vehicle to other vehicle manufacturers to allow them to comply with the ZEV mandate. See Chapter 7 for a detailed discussion of the ZEV program.

[2] The term *range-extended* PHEV is a general category based on the all-electric range of the PHEV and should not be confused with the term *extended-range electric vehicle* that General Motors uses to describe the Chevrolet Volt.

the electric grid. However, the battery is smaller than that in a BEV, and the vehicle has an onboard ICE fueled by gasoline or diesel fuel that is able to charge the battery during a trip. Although extended trips fueled only by electricity are not practical, the vehicle has a total range comparable with that of a conventional vehicle because of the onboard ICE. The 2014 Chevrolet Volt with an AER of 38 miles (DOE/EPA 2014d) is the best-selling example, and the 2014 Ford Energi models (Fusion Energi and CMax Energi) that have AERs of 20 miles are other prominent examples. The AERs are comparable to the average daily driving distance in the United States.

The consequence of eliminating the range restrictions of a limited-range BEV is that the added ICE uses petroleum and produces tailpipe emissions. Although the ICE can be operated to maximize efficiency and minimize emissions, the fraction of miles traveled propelled by electricity depends on how willing and able a driver is to recharge the battery during a trip longer than the AER. On the basis of data collected by DOE through its EV Project, early adopters of the Chevrolet Volt appear to be very motivated to minimize their use of the ICE engine by charging more frequently and logging more electric miles per day than Nissan Leaf drivers (Schey 2013). Blanco (2014) reported that 63 percent of all miles traveled by the Chevrolet Volt are fueled by electricity.

Finding: The Chevrolet Volt demonstrates that if they become widely adopted, range-extended PHEVs with AERs comparable to or greater than the average U.S. travel distance offer the possibility of significant U.S. petroleum and emission reductions without range limitations.

Type 4: Minimal Plug-in Hybrid Electric Vehicles

Minimal PHEVs are PEVs whose small batteries can be initially charged from the electric grid to provide electric propulsion for an AER that is much less than the average daily travel distance for the U.S. driver. Among many examples, the 2014 Plug-in Toyota Prius is a minimal PHEV in that its AER is only 6 miles (DOE/EPA 2014e). It is an extreme example of a car that is designed for minimum compliance with regulations rather than to give good electric-drive performance. Minimal PHEVs allow a manufacturer to comply with regulations for obtaining PEV emission credits without the expense of designing and producing a car that is optimized for using electricity instead of petroleum. They allow their drivers to comply with requirements for high-occupancy-vehicle (HOV) lane access whether or not they bother to charge from the grid (CCSE 2014). As might be expected, driver usage surveys of Plug-in Prius drivers show that a substantial fraction do not regularly charge their vehicles (Chernicoff 2014). Minimal PHEVs are essentially HEVs.

Finding: Minimal PHEVs with AERs much shorter than the average daily driving distance in the United States are essentially HEVs.

Recommendation: Minimal PHEVs should be treated as HEVs with respect to financial rebates, HOV access, and other incentives to encourage PEV adoption.

HIGH-ENERGY BATTERIES

The capacity, weight, and volume of the high-energy battery in a PEV largely determine its range, performance, and cost relative to an HEV or an ICE vehicle. This section summarizes the energy densities with respect to weight and volume that have been achieved with battery chemistries so far and considers possible improvements, despite the difficulty of precisely predicting future developments. Differences in current battery geometries and cooling strategies are discussed, along with the associated uncertainties about long-term battery durability.

Energy Density and Battery Chemistry

The battery in a PEV is the counterpart to the fuel tank for an ICE vehicle. Electric energy from the electric grid is stored in the battery until it is needed by the electric motor to turn the wheels. The more energy stored in the battery, measured in kilowatt-hours (kWh), the longer the vehicle's AER. An 80 kWh battery can propel a vehicle twice as far as can a 40 kWh battery when the same vehicle is driven in the same way, just as 20 gallons of gasoline can provide the energy to propel an ICE vehicle twice as far as 10 gallons of gasoline. The nominal battery capacities for the PEVs in Table 2-1 are what the batteries can store as their state of charge (SOC) goes from fully discharged (SOC of 0 percent) to fully charged (SOC of 100 percent). Vehicle manufacturers use electronics to restrict how fully a battery can be charged and how far the vehicle is able to deplete the charge in its battery. They make different choices for the usable capacity of their vehicle batteries because it is known that this factor affects the degradation of the battery over time, even though the degradation has yet to be fully characterized or understood.

A battery's energy density (see Figure 2-1) determines the mass and volume of the battery necessary to store the energy that a PEV requires. The vertical axis in Figure 2-1 is the energy storage capacity per unit volume (Wh/L), and the horizontal axis is the energy storage capacity per unit mass (Wh/kg). Lead acid batteries have a relatively small energy density, even though they provide starting, lighting, and ignition for essentially all the ICE vehicles around the world. The Toyota Prius was the first mass-produced vehicle to use nickel-metal hydride (NiMH) batteries. Such batteries have about twice the energy density of lead acid batteries, and they proved to be very reliable when they were used in all the early HEVs. However, there seems to be no prospect for the large increases in energy density that would be required to make them attractive for use in PEVs. Lithium-ion batteries were invented in the 1970s (Goodenough and Mizushima 1981) and mass produced for the first time by Sony for laptop computers in 1991 (Yoshino 2012). In the following two decades, lithium-

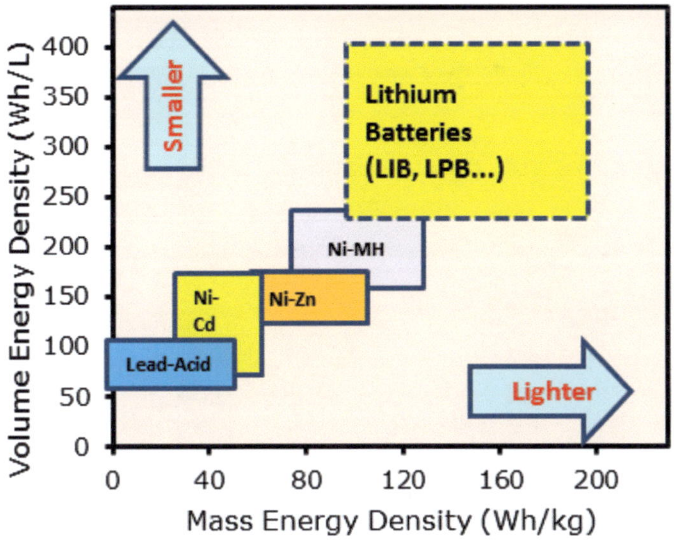

FIGURE 2-1 The volume energy density and the mass energy density for various battery types. NOTE: LIB, lithium-ion battery; LPB, lithium-polymer battery; Ni-Cd, nickel cadmium; Ni-MH, nickel-metal hydride; Ni-Zn, nickel zinc; Wh/kg, watt-hour per kilogram; Wh/L, watt-hour per liter. SOURCE: Amine (2010).

ion batteries took over the small electronics market in such devices as laptop computers and cell phones. In recent years, they have also become the battery of choice for PEVs and for new HEV models.

An electrically powered vehicle needs only about one quarter of the stored energy that an ICE vehicle needs to deliver the same energy to turn the wheels. Most of the energy that combustion releases from the fuel within an ICE is wasted as heat that is dissipated through the radiator and exhaust. The large efficiency advantage of the PEV, however, is more than overcome by the much smaller energy density in a charged battery compared with the energy density of gasoline. The result is that PEV batteries now weigh much more and occupy a much larger volume than a tank filled with gasoline. For example, the 85 kWh battery in a Tesla Model S, the largest production vehicle battery so far, weighs about 1,500 lb[3] (Tesla 2014a). Delivering the same energy to the wheels of an ICE vehicle requires the combustion of slightly less than 9 gallons of gasoline, which weighs about 54 lb.

The increased weight (about that of seven extra passengers) reduces the acceleration and the range that would otherwise be realized, although the powerful motor in the Model S overcomes the acceleration problem. Accommodating large, heavy batteries makes it difficult to use an ICE or HEV platform for an electric vehicle. A vehicle designed from its beginning to have electric propulsion has more options. The Model S, for example, was designed with a battery compartment under the vehicle's entire floor board so that the heavy batteries are used to keep the vehicle's center of gravity low to improve handling.

[3] The estimate is based on Tesla's reported energy density for the Model S battery of 121 Wh/kg (Tesla 2014a).

The lithium-ion batteries in vehicles differ in the chemistries and materials that are used and in the energy densities achieved (Table 2-2). In a lithium-ion battery (see Figure 2-2), the positive lithium-ions flow between the anode and the cathode within the electrolyte, as do electrons in an external circuit connected between the anode and cathode. The cathodes used are described using chemical formulae that provide their composition. All anodes but one are carbon. All PEV batteries use an organic solution of $LiPF_6$ as the electrolyte.

The committee notes that the design of a vehicle battery is related not only to the battery chemistry but also to the power and energy requirements of the various applications. For example, PHEVs require more power than BEVs; thus, BEVs can use thicker, cheaper electrodes. Furthermore, PHEV batteries must be cycled more frequently than BEV batteries, so PHEV batteries tend to use a smaller portion of the nominal battery capacity. Those two facts affect the battery structure and cost per kilowatt-hour and are taken into account in various analyses of PEV battery costs (Daniel 2014; Sakti et al. 2014) and in the EPA/NHTSA analysis that informed the committee's analysis of battery costs as discussed below.

Projected Energy Density Increases and Possible New Battery Chemistries

Lithium-ion batteries with increased energy density are naturally the subject of research and development efforts. It is difficult to predict success or its timing, but three approaches that are being pursued are worthy of mention.

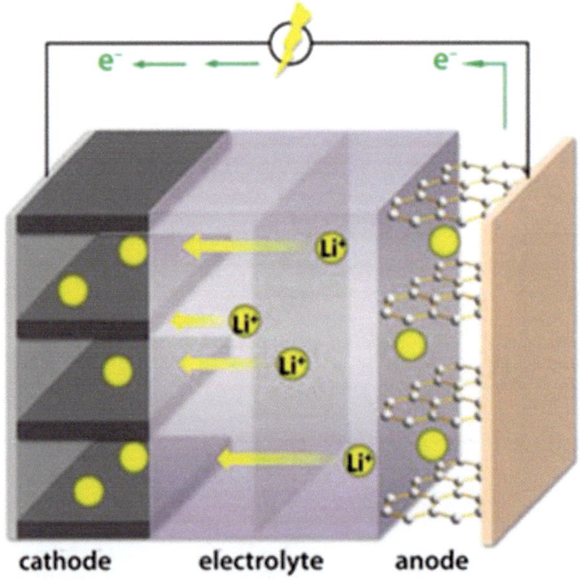

FIGURE 2-2 Representation of a lithium-ion battery that shows lithium ions traveling between the anode and the cathode and electrons traveling through the external circuit to produce an electric current. SOURCE: Kam and Doeff (2012).

TABLE 2-2 Properties of Lithium-Ion Batteries in Four Plug-in Electric Vehicles on the U.S. Market

PEV	Cathode	Anode	Supplier	Cell Type	No. of Cells	Energy (kWh)	Power (kW)
Tesla Model S	NCA = $LiNi_{0.8}Co_{0.15}Al_{0.05}O_2$	Carbon	Panasonic	Cylindrical	~8,000	85	270
Chevrolet Volt	LMO = $LiMn_2O_4$	Carbon	LG Chem	Prismatic	288	16.5	111
Nissan Leaf	LMO = $LiMn_2O_4$	Carbon	Nissan/NEC	Prismatic	192	24	90
Honda Fit	NMC = $LiNi_{1/3}Mn_{1/3}Co_{1/3}O_2$	$Li_4Ti_5O_{12}$	Toshiba	Prismatic	432	20	92

NOTE: Al, aluminum; Co, cobalt; kWh, kilowatt-hour; Li, lithium; LMO, lithium manganese oxide; Mn, manganese; NCA, nickel cobalt aluminum oxide; NMC, nickel manganese cobalt oxide; Ni, nickel; O, oxygen; Ti, titanium.

- Increasing the number of lithium atoms in a layered cathode structure has been shown in the laboratory to increase the energy density (Julien et al. 2014).
- Developing electrolytes that can operate at 4.8 V rather than 4.2 V would increase the energy density (Pham et al. 2014).
- Replacing the carbon anode with one that includes silicon would improve the energy density (Ge et al. 2013). Theoretically, a pure silicon anode would have an energy density 10 times that of a pure carbon anode. However, pure silicon anodes are not practical because they crumble during a charging cycle, being unable to withstand having their volume changed by more than a factor of three. Mixtures of silicon and carbon with appropriate binders might minimize the volume change and yet provide an increased energy density.

The committee estimates that although there can be no guarantee, as much as a twofold increase in energy density could come from some combination of the three approaches within the next decade. Such an increase would allow an important reduction in the volume and weight of high-energy batteries. Most important, however, the cost per kilowatt-hour needs to decrease; a battery having twice the energy density at twice the cost would not make PEVs any more affordable. Nonetheless, even with such an improvement, battery energy densities would still be much smaller than the energy density of gasoline.

On a longer time scale, other battery chemistries could significantly increase the energy density. The theoretical energy density for a lithium-air battery is 5,200 Wh/kg (Rahman et al. 2014), which is comparable to that of gasoline. Such a battery uses oxygen from air and therefore does not need to store an oxidizer. PolyPlus (2009) claims to have a battery capable of 700 Wh/kg and expects to produce a rechargeable battery with a higher energy density. Another promising approach is the development of a high-energy density lithium-sulfur battery. Sion Power, the recipient of substantial Advanced Research Projects Agency-Energy (ARPA-E) funding, claims that "over 600 Wh/kg . . . and 600 Wh/L in energy density are achievable in the near future" (Sion Power 2014). Substantial challenges remain for both lithium-air and lithium-sulfur batteries, however, particularly in producing batteries that survive frequent recharging, so it is difficult to predict if and when batteries with much higher energy densities will be available.

Finding: Affordable batteries with higher energy densities and longer useful lives could greatly increase the all-electric range and presumably increase the adoption rate for PEVs.

Finding: Although there can be no guarantee, as much as a twofold increase in energy density from present values of 100-150 Wh/kg could come from some combination of current research efforts within the next decade.

Finding: Battery research is critical because more practical vehicle batteries that have higher energy densities and longer life are needed to address important concerns about battery range and durability.

Battery Geometry, Cooling, and Durability

Just as there is no consensus on what is the best lithium-ion battery chemistry, there is also no consensus on what is the most stable or most economical battery geometry or on how much the battery temperature should be regulated for the sake of battery longevity. As more PEVs are driven, the early adopters are essentially testing both the various battery chemistries and the battery temperature regulation choices under real-world conditions that are hard to duplicate in laboratories.

Tesla connects many thousands of small cylindrical cells, each having the same physical shape and size as those that are commonly used in computer batteries, thereby profiting from the extensive manufacturing experience for cells with this geometry. All other manufacturers use many fewer but much larger cells in so-called prismatic or pouch geometries. A Nissan Leaf air-cools its batteries, while the Chevrolet Volt and the Tesla Model S use a liquid system and heat exchangers to regulate battery temperature. Over the next several years, the real-world experience reported by early adopters should make clear the advantages and disadvantages of each strategy.

Concerns about the durability and performance of the current lithium-ion batteries at extremely high and low temperatures could be a barrier to PEV adoption, depending on the durability observed as more vehicles are driven for longer times (Steffke et al. 2008). One study that evaluated a PHEV with a 20 kWh battery showed that a hot climate accelerates the normal degradation of battery capacity with time (see Figure 2-3) (Pesaran et al. 2013). Reports on shorter battery life for Nissan Leafs in Arizona seem consistent with that observation (Gordon-Bloomfield 2013). As a result, Nissan has tested new battery pack designs to address the observed problem (Gordon-Bloomfield 2013), and press reports of the increased rate of battery deterioration have not continued. However, it is not clear whether the problem has been solved. Although Figure 2-3 illustrates preliminary results of studying the effect of temperature on battery capacity, battery life depends also on cycling at various depths of charge, rate of charge and discharge, and likely many other variables besides temperature. Only long-term experience in hot climates will establish whether some manufacturers must improve battery temperature regulation, use different battery chemistries, or restrict sales in hot climates.

ICE vehicle manufacturers have a good understanding of how long their products will perform, and this knowledge allows them to predict warranty costs. PEV manufacturers are still learning about battery longevity. As more PEVs enter the market, vehicle manufacturers have the chance to experiment with various warranties and battery maintenance contracts as they look for affordable ways to reassure and share risk among consumers that use these vehicles under real-world conditions. Vehicle leasing is becoming more popular and promoted by some manufacturers partly because this option allows a consumer to avoid long-term liability for a battery if over time the battery performance degrades below an acceptable level.

Finding: Concerns about the durability and performance of the current lithium-ion batteries at extremely high and low temperatures could be a barrier to PEV adoption, depending on the durability observed as more vehicles are driven for longer times.

RELATIVE COSTS OF PLUG-IN ELECTRIC AND ICE VEHICLES

Studies of current and projected costs of high-energy batteries and nonbattery components (EPA/NHTSA 2012) suggest that the difference in cost of producing a PEV and an ICE vehicle is (and will be) primarily due to the cost of the high-energy battery. Those studies are part of the regulatory analysis performed by EPA and the National Highway Traffic Safety Administration (NHTSA) for the recent 2017-2025 combined CAFE-GHG standards for light-duty vehicles. The comprehensive regulatory analysis includes vehicle-simulation modeling and detailed component cost analysis (cost teardown studies) performed by external consultants to determine cost and effectiveness of a wide range of technologies, including conventional ICE vehicles, HEVs, and PEVs. Thus, for its assessment, the committee relied on the CAFE-GHG regulatory analysis (EPA/NHTSA 2012), as well as on presentations from vehicle manufacturers, suppliers, and market analysts (Tamor 2012; Ward 2013; Woodard 2012; Sriramulu

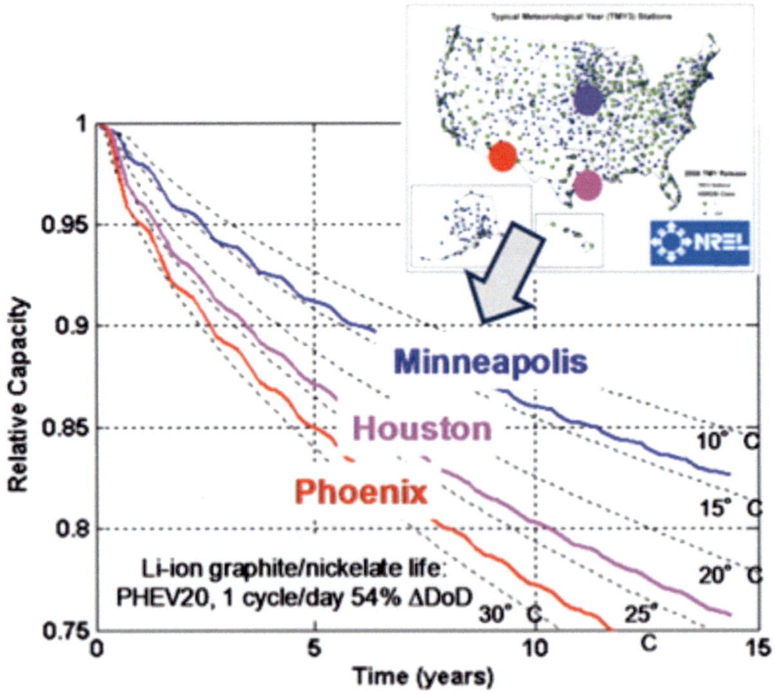

FIGURE 2-3 Effect of ambient temperature on battery capacity on a 20 kWh battery in a PHEV. NOTE: DoD, depth of discharge; PHEV, plug-in hybrid electric vehicle. SOURCE: Pesaran et al. (2013).

and Barnett 2013; Anderman 2014), because a detailed independent cost analysis was beyond its scope and resources. The committee also reviewed the cost information provided in *Transitions to Alternative Vehicles and Fuels* (NRC 2013). That committee estimated battery costs by assuming that future costs for Li-ion cells for vehicles would follow a similar, although slower, cost reduction trajectory as that experienced by Li-ion 18650 cells. Although cost projections were somewhat similar, this report makes use of the recent extensive analysis done specifically for the costs of vehicle Li-ion batteries. Costs of the batteries and nonbattery components are discussed below; vehicle price and cost of ownership are discussed further in Chapter 7.

Lithium-Ion Battery Costs

A high-energy battery costs much more than a sheet-metal gasoline tank. Studies of current and projected battery costs are summarized here to estimate the magnitude of the cost differential and whether it is likely to continue. *Cost* refers to what a vehicle manufacturer would pay a supplier, which is known as the direct manufacturing cost (DMC) (EPA/NHTSA 2012). What a consumer would pay for a battery (the retail price equivalent) is expected in the automotive industry to be about 50 percent more than what a vehicle manufacturer would pay (NRC 2011). Large price fluctuations must be expected until battery supply and demand for PEVs becomes more predictable. Until then, the price will likely depend strongly on the availability of unused battery production capacity and a manufacturer's desire to be perceived as a technology leader. It might further depend on the willingness of the vehicle manufacturer to set a price that allows it to gain a market share for its vehicles.

Unfortunately, there are no definitive studies of battery costs from battery manufacturers given their need to protect proprietary information. The range of cost projections from studies of current and future battery costs is considerable. An additional complication is that vehicle manufacturers make different choices on how much of the total capacity of a battery is made available for use; GM uses about 70 percent of the nominal capacity, and Nissan uses about 90 percent (see Table 2-1).[4] To allow comparisons, the committee converted study results to be the projected costs per kilowatt-hour of the total battery capacity rather than the available battery capacity. The costs estimated below are for complete battery packs, excluding any cooling system.

- A 2012 Argonne National Lab study projected costs to be between \$251 and \$280/kWh for a battery pack produced in 2020 converted to 2012 dollars (Nelson et al. 2011).

[4] The values cited seem appropriate given that PHEV batteries could be cycled more times per trip than BEV batteries and that using a smaller portion of the nominal capacity increases battery life.

TABLE 2-3 Estimates of Dollars per Kilowatt-hour for a 25 kWh Battery

Year	Manufacturing Volume (packs/year)	Cell Materials ($/kWh)	Cell Price ($/kWh)	Pack Price ($/kWh)
2013	25,000	110-150	275-325	400-500
2016	50,000	90-130	185-230	275-350
2020	100,000	85-110	140-190	225-275

SOURCE: Based on data from Anderman (2014).

- TIAX projected that direct material and direct labor costs would amount to $310/kWh for an annual production volume of 300,000, a large number compared with U.S. PEV sales to date (Sriramulu and Barnett 2013).
- DOE has estimated a current cost of $240/kWh (Howell 2013).[5]
- EPA/NHTSA (2012) projected $540, $346, and $277/kWh for a PHEV40 with a 16 kWh battery pack in 2017, 2020, and 2025, for an annual volume of 400,000.
- A 2011 McKinsey study estimated the costs to be $350 to $420/kWh; it predicted that these costs would drop to about $140/kWh by 2020 and $112/kWh by 2025 (Hensley et al. 2012).
- Anderman (2012) predicted that the cost for a 24 kWh battery pack in the 2015 time frame in volumes of 100,000 units would be $340 to $450/kWh.
- Anderman (2014) provided estimates of dollars per kilowatt-hour for a 25 kWh battery (see Table 2-3).

An attempt has been made to convert study results to cost per kilowatt-hour of the total energy that can be stored in the battery and to 2013 dollars (see Table 2-4).

The range of estimates in the current studies show that current costs are difficult to obtain and that the future projections are even more difficult, requiring, for example, an estimate of how many PEVs will be purchased. For the purposes of this report, the committee decided to use the $500/kWh as the current cost of the lithium-ion battery pack and about $250/kWh as the cost in about 10 years. Thus, at $500/kWh, the DMC of the Tesla battery would be $42,500, the DMC of the Leaf battery would be $12,000, the DMC of the Volt battery would be $8,250, and the DMC of the Plug-in Prius battery would be $2,200.

Figure 2-4 shows the decrease in costs of the Li-ion battery cell over the last 13 years and illustrates how Tesla has profited from the reduced prices for the small cell package used to power consumer electronics. The recent prices shown for Li-ion batteries in Figure 2-4 ($400/kWh) correspond to a cost of about $270/kWh if the assumption mentioned earlier is used that price is 1.5 times the cost. Some care is required in deducing cost from prices in recent years because battery manufacturers might be reducing prices to cope with having more production capability than demand. Some reports suggest that Tesla is paying much less for batteries from Panasonic. In addition, Tesla has announced plans to build a $5 billion battery factory and has stated that it believes it can substatially reduce battery costs (Trefis Team 2014). The committee does not have any information about how the cost reductions will be achieved, but the factory investment appears to be a strong indication that Tesla is confident that it can build high-energy batteries more economically than has so far been possible.

Finding: It is not possible to determine a completely reliable projection of future battery cost. However, given the available data, the committee assumed for this report a battery pack cost of $500/kWh in 2013 and a 50 percent lower cost in about 10 years.

Finding: The high cost of high-energy batteries is primarily responsible for the higher initial cost of PEVs compared with HEVs and ICE vehicles and is a barrier to PEV adoption.

Finding: Even if the higher initial battery cost drops as predicted over the next 10 years, battery cost will remain a barrier to PEV adoption.

Nonbattery Costs

An ICE vehicle includes an ICE, a radiator, a transmission, and an oil system. A BEV has instead an electric motor; power electronics that convert the direct current (dc) power from the battery to the alternating current (ac) power needed to drive the electric motor; and electronics needed to charge the battery. A PHEV includes both sets of components. The nonbattery costs of the PEV are primarily attributable to the power electronic controls and the electric motor and generators. The committee reviewed and accepted the estimates for nonbattery costs from the EPA/NHTSA (2012) study that was used to evaluate CAFE standards because it found that the cost analysis performed by the agencies was thorough and comprehensive.

The simplicity of a BEV compared with an ICE vehicle makes it somewhat surprising that the EPA/NHTSA (2012) study estimates that the direct manufacturing cost of the nonbattery components for a BEV with a range of 75 miles is about $1,255 higher than the cost of the ICE power-train components it replaces. The increased cost includes $3,810

[5] A current cost estimated to be $300/kWh becomes $240/kWh for the total battery capacity, assuming that the original estimate was for an 80 percent utilization of the battery.

TABLE 2-4 Summary of Estimated Costs of Total Energy from Various Sources (2013 U.S.$/kWh)

	Year				
Source	Current[a]	2017	2020	2022	2025
Argonne					
2000	250-706	—	—	—	—
2012	—	—	50 kW = 336 100 kW = 404	—	—
TIAX 2013	310	—	—	—	—
DOE 2013	300	—	—	125	—
EPA/NHTSA 2012	—	540	346	—	277
McKinsey 2011	350-420	—	140	—	112
Anderman					
2012	340-450	—	—	—	—
2014	400-500	—	220-275	—	—

[a] Current as defined in the respective studies.

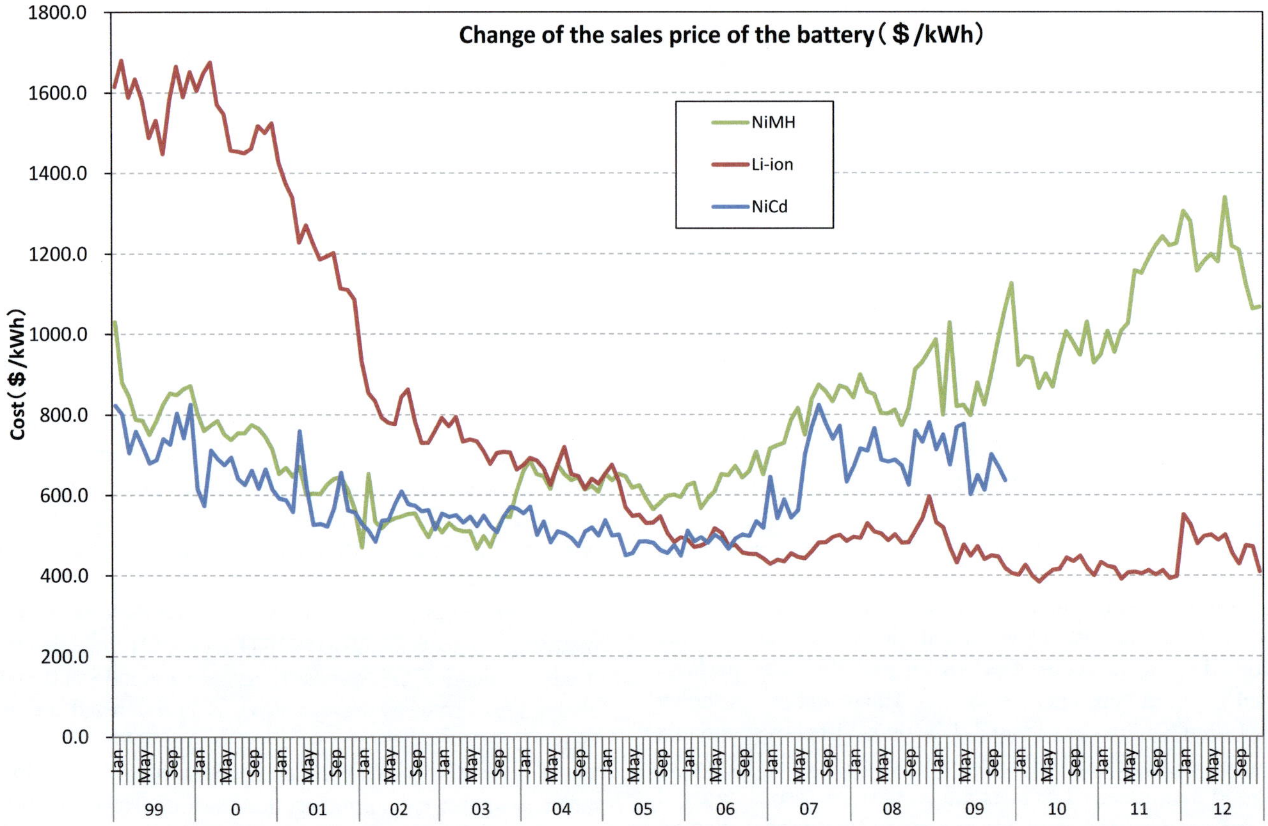

FIGURE 2-4 Change in the sales price of NiMH, Li-ion, and NiCd battery cells from 1999 to 2012. Prices are shown in 2012 dollars. The graph is based on data from a production survey conducted by the Ministry of Economy, Trade, and Industry, Japan. NOTE: kWh, kilowatt-hour; Li-ion, lithium ion; NiCd, nickel cadmium; NiMH, nickel-metal hydride. SOURCE: Maruyama (2013).

for the electric motor, inverter, high-voltage wiring, and improvements in the climate-control system. Those component costs are partially offset by the elimination of the ICE, transmission, and related components, which account for a savings of $2,555 in direct manufacturing costs (EPA/NHTSA 2012). The EPA/NHTSA estimates that the nonbattery costs in 2025 will drop to 80 percent of their 2012 costs. However, even if the cost reduction is less, the cost of the high-energy battery will still account for most of the difference in cost between a BEV and an ICE vehicle.

Because a PHEV has both an electric drive and an ICE, it has a higher nonbattery cost. The same study evaluated a PHEV with a 40-mile AER and concluded that a PHEV has nonbattery cost that is $3,700 higher than the nonbattery cost of an ICE vehicle. Multiplying by 1.5 increases the price to the consumer to $5,550 beyond the price of the battery.

A dramatic reduction in the price of power inverters could potentially come from the replacement of silicon-based semiconductors by wide bandgap materials, such as SiC and GaAs, that would enable faster switching and lower resistance to improve the inverter efficiency. Those materials operate at much higher temperatures than the silicon used in today's power electronics, and that characteristic would make cooling easier and thereby reduce the size of the power electronics package and possibly simplify the heat exchangers (ORNL 2012). However, when such technology will be far enough along to come to market is difficult to predict.

Finding: Because power electronics and large electric motors are new to the automotive industry, nonbattery costs will likely drop substantially as new models come to market.

VEHICLE CHARGING AND CHARGING OPTIONS

Charging a PEV is analogous to filling a conventional vehicle's fuel tank with gasoline. A gasoline-powered vehicle is attached to a pump that sends gasoline through a hose into the fuel tank. A typical flow rate of 8 gal/min, for example, means that typical gasoline tanks with capacities of 10 to 20 gal will be filled in a few minutes. Similarly, a PEV is plugged into the electric grid so that electricity can flow through wires into the battery. An energy flow rate of 6.6 kW, for example, would fill an empty battery with a usable capacity of 21 kWh in about 4 hours.

The maximum charging rate for residential charging is limited by the size of the charger in the vehicle that changes ac electricity into dc electricity. A fully discharged battery initially charges at the maximum rate that the onboard charger can manage and then charges more slowly as the battery nears capacity. Thus, a vehicle battery does not charge at a constant rate, and that is why it takes about 4 hours to fill a 21 kWh battery at 6.6 kW. For DC fast charging (discussed below), the component that changes ac to dc is outside the vehicle and is governed by control signals from the vehicle. Regulating the charging rate is necessary to ensure safety and to protect battery life. Although increasing the charging rate with high-power chargers shortens the time needed to charge a vehicle's battery, an important technical issue now being researched is the extent to which faster charging at high power hastens the normal aging of a battery (Francfort 2013).

The "pressure" with which an electric circuit in a home or business can force electricity through wires into some device is measured in volts (V). The amount of electricity flowing through various devices, the electric current, is measured in amperes (A). The product of the two is the power in watts (W). Every circuit delivering electricity has a circuit breaker or fuse that keeps the flow of electricity from exceeding the amperes that the circuit can safely provide. For example, a 2014 Nissan Leaf is capable of accepting no more than 30 A of electric current when it is connected to a 240 V electric circuit, so its maximum power consumption is 7.2 kW. The vehicle will not accept more current or power even if the circuit is able to provide it. The circuit is protected by a 40 A circuit breaker, resulting in what is referred to as a 240 V, 40 A service.

As recommended by the National Electrical Code (NEC), an apparatus known as the electric vehicle supply equipment (EVSE) is always connected between the charging circuit and the vehicle to protect the people and the vehicle during charging. The purpose of the EVSE is to create two-way communication between vehicle and charger before and during charging to detect any anomalies that might affect safety or the equipment (Rawson and Kateley 1998). The NEC (2008) defines the EVSE as "the conductors, including the ungrounded, grounded, and equipment grounding conductors and the electric vehicle connectors, attachment plugs, and all other fittings, devices, power outlets or apparatus installed specifically for the purpose of delivering energy from the premise's wiring to the electric vehicle" (Section 625.2). Its ground fault interrupters—like those in bathrooms and kitchens—are safety devices that can detect when a small electric current from the circuit has "gone missing" and disconnect the electric circuit and the current flow before anyone is injured. Furthermore, the EVSE is able to communicate with a vehicle to ensure that no current is provided before the vehicle is connected. The EVSE for slow charging via 120 V is typically a portable device that can be carried in the vehicle for possible use at remote locations. The EVSE for normal 240 V charging is typically mounted on a garage wall or on a purpose-built column. Fast chargers that use high dc voltages have the EVSE built into the substantial charger that is required.

For EVSEs connected to the single phase 120 V ac or the split-phase 240 V ac circuits that are commonly available in U.S. homes and workplaces, a plug wired to the EVSE connects to a socket on the vehicle. The circuit breaker or fuse sets the maximum current that the EVSE can provide, although individual vehicles will typically accept less current. In the United States, there is one standard plug that is used to charge vehicles from the normal 120 V and 240 V circuits found in residences, the SAE J1772 standard (SAE 2012).

This interchangeability removes what otherwise could be a substantial barrier to the adoption of PEVs. However, for faster charging options, fast chargers are being installed that have one or more of three incompatible plugs and protocols described below.

AC Level 1 Charging

Most electric devices in the United States (for example, lamps, small air conditioners, and computers) are plugged into single-phase 120 V ac electric circuits accessed via the wall sockets present in essentially every room of every building. Circuit breakers or fuses switch off the electricity if the current flowing through the circuit exceeds 15 to 20 A to prevent fires and other damage to the circuits.

AC level 1 charging standard is for an EVSE that plugs into a 120 V wall plug (Figure 2-5) and delivers up to 12 A to a SAE J1772 plug (Figure 2-6), which connects with a socket in the car. Most PEVs today have an onboard charger that changes the ac current into the dc current that charges the battery. The charger is able to accept only up to 12 A from the EVSE and transfer energy at a rate of up to 1.4 kW. Much like the largest window air conditioners that can be plugged into a 120 V circuit, the vehicle that is charging must typically be the only device drawing current from the circuit to avoid exceeding the maximum current that the circuit breaker or fuse will allow the circuit to provide.

PEVs are typically sold with a small and portable EVSE that can be carried in the car to allow AC level 1 charging from ubiquitous 120 V wall receptacles. A deficiency of the standard is that the portable EVSE is not secured to either the 120 V socket or to the vehicle to deter EVSE theft or vandalism. AC level 1 charging with this EVSE is the only charging option typically needed or available for the minimal PHEVs. Each hour of charging typically provides an additional electric range of about 4 to 5 miles, depending on the vehicle. For a range-extended PHEV, such as a Chevrolet Volt, some drivers use only AC level 1 charging, while others prefer to charge about twice as fast using the AC level 2 charging that is discussed below.

For charging the fully depleted batteries of PEVs with large batteries, AC level 1 charging is too slow to be the primary charging method because charging times could be longer than the time that a car is parked at the home or workplace. For example, with an AC level 1 charger, the nominal time for fully charging the usable 21 kWh capacity of a

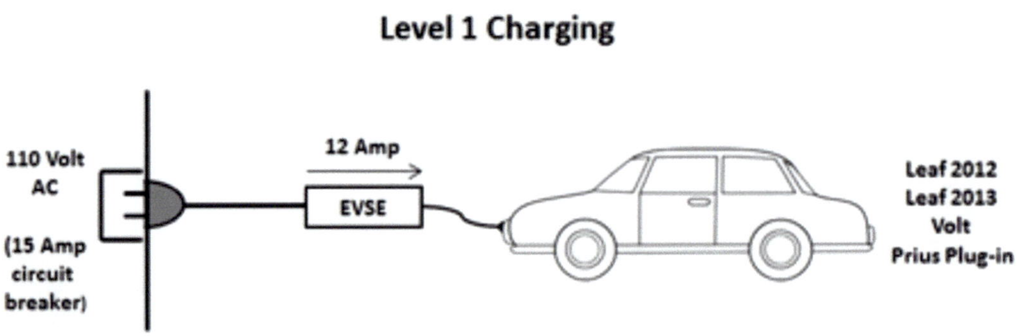

FIGURE 2-5 For AC level 1, a vehicle is plugged into a single-phase 120 V electric socket through a portable safety device called an electric vehicle supply equipment (EVSE).

FIGURE 2-6 The SAE J1772 plug that connects all PEVs to AC level 1 and level 2 is an agreed-on universal standard for 120 V and 240 V ac charging. SOURCE: © Michael Hicks, licensed under Creative Commons 2.0 (CC-BY-2.0).

Nissan Leaf battery is more than 17 hours, and the nominal time to fully charge the 85 kWh battery of a Tesla Model S is more than 61 hours. However, AC level 1 charging could be useful in some cases to merely extend the range of those BEVs by a few miles if that is all that is needed.

AC Level 2 Charging

AC level 2 charging uses a 240 V, split-phase ac circuit (Figure 2-7). Such circuits are available in essentially all homes and workplaces and are used by electric dryers, electric stoves and ovens, and large air conditioners. Since 2009, the AC level 2 standard allows up to 80 A of current to be delivered for an energy transfer rate of 19 kW, although the wiring in many houses will have trouble delivering that much current, and only a long-range BEV is capable of accepting it. A Chevrolet Volt and a 2014 Nissan Leaf are able to accept a maximum of 12 A or 30 A, respectively, which corresponds to energy being transferred at maximum rates of 3.3 and 7.2 kW, respectively. As noted, the 240 V EVSE for AC level 2 charging is typically wall-mounted in a garage or on a post next to a parking spot, and in the United States, it is connected to the vehicle through the same SAE J1772 plug (Figure 2-6) used for AC level 1 charging.

The 85 kWh battery of the Tesla Model S, much larger than the battery in any other PEV, is the only vehicle battery so far that can accept the highest rated current and power from an AC level 2 charging system. The normal home charging recommendation is to deliver 40 A and nearly 10 kW to a "single" charger installed in the Tesla Model S. If enough current is available in a home, a "double" charger can instead be installed in the car to accept 80 A and 19 kW power for much faster charging. With that option, Tesla advertises that the car can travel an additional 58 miles for each hour of charging (Tesla 2014b). For emergency use, the Tesla Model S also supplies a portable EVSE with adapters that allow it to be charged using most of the common 240 V wall sockets that deliver 24 or 40 A to electric dryers, stoves, and air conditioners.

DC Fast Charging

Faster charging is generally carried out by supplying a high dc voltage directly to the battery. In this case, the charger that turns the ac electricity available from the grid into the dc electricity required to charge the battery is located in the EVSE rather than within the car. Such charging is only useful for limited-range and long-range BEVs, such as the Nissan Leaf and the Tesla Model S, and only BEVs are typically able to accept fast charging.

A proliferation of incompatible connector (and protocol) standards are used for the DC fast chargers. Four options are being offered worldwide (Figure 2-8), three of which are becoming increasingly available in the United States.

All fast chargers installed in the United States so far are CHAdeMO chargers with the exception of the Tesla superchargers.[6] The Nissan Leaf accepts a CHAdeMO plug (Figure 2-8A), which provides the high voltage dc and control signals to the vehicle. A 44 kW CHAdeMO charger can charge a Nissan Leaf to 80 percent of its capacity in 30 minutes (see Figure 2-9).

The Tesla Model S accepts a proprietary fast-charging plug (Figure 2-8C), and charges are free at Tesla superchargers for models with an 85 kWh battery (that is, such charging is included in the purchase price of the vehicle). Existing 90 kW superchargers are being upgraded to 120 kW so that

[6] In October 2014, the total number of CHAdeMO chargers worldwide was 4,180, with the following breakdown: Japan, 2,129; Europe, 1,327; United States, 700; and other, 24 (CHAdeMO 2014).

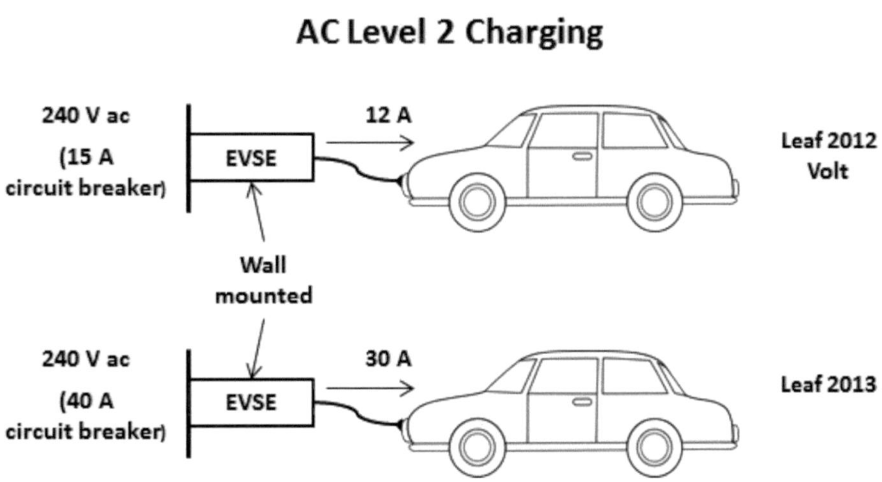

FIGURE 2-7 For AC level 2 charging, a vehicle is plugged into a split-phase 240 V electric circuit like those used by electric dryers, stoves, and large air conditioners through a wall- or post-mounted safety device called an electric vehicle supply equipment (EVSE).

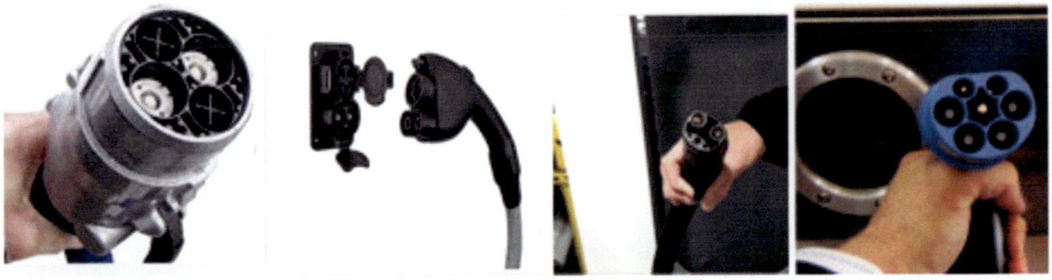

A B C D

FIGURE 2-8 Four plugs and control protocols are now being used for DC fast charging: (A) the CHAdeMO plug that is used for the Nissan Leaf; (B) the SAE J1772 combo standard that is used on the BMW i3 and the Chevrolet Spark. The upper part of the connector is the same as the SAE J1772 plug that is used universally in the United States for AC level 2 charging (see Figure 2-6); (C) the proprietary Tesla plug that is used for the Tesla supercharger network; and (D) the Mennekes plug recently adopted by the European Union for use in Europe. SOURCE: (A) © C-Car-Tom, licensed under Creative Commons 3.0 (CC-BY-3.0); (B) SAE (2012), reprinted with permission from SAE J1772 Feb2012 © 2012 SAE International; and (D) © loremo, licensed under Creative Commons 2.0 (CC-BY-2.0).

FIGURE 2-9 DC fast charging a Nissan Leaf. DC fast charging is able to charge a Nissan Leaf battery to 80 percent capacity in 30 min. The charge would typically allow a 2014 Nissan Leaf to travel about 67 miles. SOURCE: Copyright © 2010 by the eVgo Network, licensed under Creative Commons 2.0 (CC-BY-2.0).

the battery can be charged to 50 percent of its capacity in as little as 20 minutes. The announced goal is to install 250 units so that 98 percent of U.S. drivers are within 100 miles of a supercharger by the end of 2015 (Tesla 2014c). The locations of the superchargers are shown in Figure 2-10. Tesla chargers will not be available to drivers of other long-range BEVs when these become available.

The SAE added a dc and a ground lead to the SAE J1772 plug universally used for AC level 2 charging (Figure 2-6) to make a J1772 combo plug (Figure 2-8B). There are almost no installed combo chargers in the United States to date and few PEVs that are able to use them. However, the Chevrolet Spark and the BMW i3 that is just becoming available in the United States use them.

The European Union recently adopted the Mennekes (Masson 2013) plug (Figure 2-8D) for its 240 V AC level 2 standard for charging rates up to 39 kW. That standard is not discussed in detail because it is not expected to be used in the United States.

The variety of DC fast-charging plugs and communication protocols seems unfortunate. For long-range BEVs, the future situation could be like having separate networks of gasoline stations for ICE vehicles made by different manufacturers. It is not a big problem now in that the Tesla Model S is the only long-range BEV able to make long trips using the proprietary network of Tesla superchargers. As other manufacturers introduce long-range BEVs, however, they might need to introduce their own charger networks to compete. The United States and proactive states like California might be able to use their influence and incentives to make it possible to fast charge any PEV at any fast-charging station. The United States could raise the issue of compatible charger designs in free trade talks with the European Union and with its trading partners in Asia.

Finding: A network of fast-charging stations is currently being completed by Tesla without the use of public funds. However, it is a proprietary network that might not be avail-

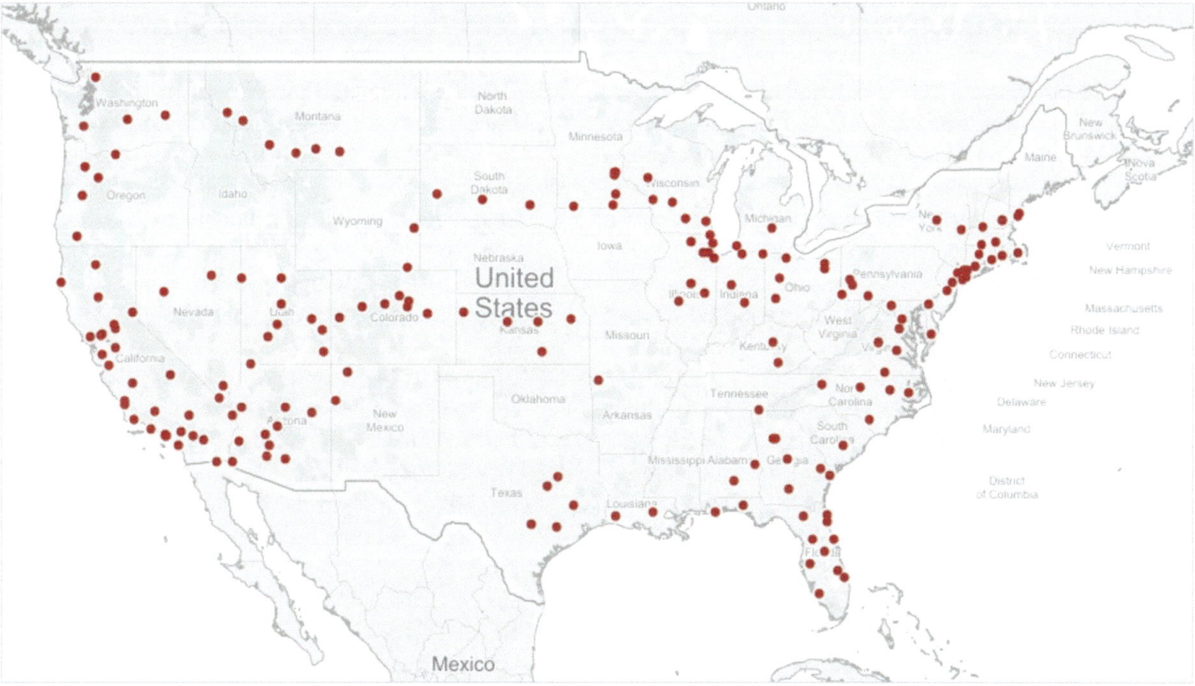

FIGURE 2-10 As of February 2015, Tesla had installed 190 units in the United States. SOURCE: DOE (2015).

able for the use of all drivers when more long-range BEVs come to market.

Finding: The various plugs and communication protocols that are used across the world for charging PEVs are a barrier to the adoption of PEVs insofar as they prevent all PEVs from being able to charge at any fast-charging station.

Recommendation: The federal government and proactive states should use their incentives and regulatory powers to (1) eliminate the proliferation of plugs and communication protocols for DC fast chargers and (2) ensure that all PEV drivers can charge their vehicles and pay at all public charging stations using a universally accepted payment method just as any ICE vehicle can be fueled at any gasoline station. The Society of Automotive Engineers, the International Electrotechnical Commission, and the Verband der Elektrotechnik—companies that formed CHAdeMO—and Tesla should be included in the deliberations on plugs and communication protocols.

Wireless Charging

So far, essentially all PEVs are charged by plugging a charging cable into the vehicle so that electricity can flow from the EVSE to the battery. The process is simple and rapid (less than a minute), and control electronics are included to enhance safety.

Wireless charging would instead transfer the energy from the grid to the vehicle by using inductive coupling between a wireless transmitter located near the vehicle and a wireless receiver attached to the vehicle (Miller et al. 2014). An alternating magnetic field produced by passing ac current through coils in the wireless transmitter would induce a voltage in the coil of the receiver. The latter currents would charge the vehicle battery. Static and dynamic wireless charging are possible.

Static wireless charging takes place when the vehicle is not moving, as described. The energy transfer is less efficient than using a charging cable, but there would be no cable to handle or keep clean. For publicly available charging, standards would be needed to make it possible to charge most PEVs with most wireless charging systems. The opportunity for theft or vandalism of the cable or EVSE is greatly reduced because the transmitter could be embedded in the parking space and controlled remotely. A safety standard to establish the acceptable levels of oscillating electromagnetic fields might also be needed.

Dynamic wireless charging is a futuristic concept that is being investigated to see if it might ever be feasible (Miller et al. 2014). The vision is that a vehicle would receive power in its wireless receiver as it passed long series of wireless transmitters, so a BEV could be refueled on long trips without stopping to refuel. However, there are many technical problems to overcome for dynamic wireless charging, one of them being a very low charging efficiency.

REFERENCES

Amine, K. 2010. "Advanced Battery Chemistries for PHEV and EV Applications." Presentation at the U.S.- China Electric Vehicle and Battery Technology Workshop, Argonne, IL, August 30-September 1. http://www.transportation.anl.gov/batteries/us_china_conference/docs/roundtable1/adv_battery_chem_amine.pdf.

Anderman, M. 2012. "Energy Storage for xEVs: Status, Trends and Challenges." Presentation at the 24th International AVL Powertrain Conference Engine and Environment, Graz, Austria, September 12.

Anderman, M. 2014. "Li-Ion Technology Evolution for xEVs: How Far and How Fast?" Presentation to the Committee on Overcoming Barriers to Electric-Vehicle Deployment, Irvine, CA, February 25.

Blanco, S. 2014. "Chevy Volt Owners Log Half a Billion Electric Miles, 2015 Production Starts." Autoblog-Green, June 19. http://green.autoblog.com/2014/06/19/chevy-volt-half-billion-miles-2015-production/.

CCSE (California Center for Sustainable Energy). 2014. February 2014 Survey Report. http://energycenter.org/clean-vehicle-rebate-project/vehicle-owner-survey/feb-2014-survey.

CHAdeMO. 2014. "CHAdeMO's Fast Charging Station in the World." http://www.chademo.com. Accessed October 1, 2014.

Chernicoff, W.P. 2014. Perspectives on Electric Vehicles from Toyota. Presentation to the Committee on Overcoming Barriers to Electric-Vehicle Deployment, Irvine, CA, February 25.

Consumer Reports. 2014. "Tesla Model S 2013- 2014 Quick Take" [video file]. http://www.consumerreports.org/cro/video-hub/cars/hybrids--alternative-fuel/tesla-model-s-20132014-quick-take/14786539001/2366240882001/. Accessed March 28, 2014.

Daniel, C. 2014. "Lithium Ion Batteries and Their Manufacturing Challenges." Presentation at NAE Frontiers of Engineering, Irvine, CA, September 11-13. http://www.naefrontiers.org/File.aspx?id=46415.

DOE (U.S. Department of Energy). 2012. 2011 Nissan Leaf -VIN 0356 Advanced Vehicle Testing—Beginning-of-Test Battery Testing Results. Vehicle Technologies Program. http://www1.eere.energy.gov/vehiclesandfuels/avta/pdfs/fsev/battery_leaf_0356.pdf.

DOE. 2013. 2013 Chevrolet Volt-VIN 3929 Advanced Vehicle Testing—Beginning-of-Test Battery Testing Results. Vehicle Technologies Program. http://www1.eere.energy.gov/vehiclesandfuels/avta/pdfs/phev/battery_volt_3929.pdf.

DOE. 2015. "Alternative Fueling Station Locator." Alternative Fuels Data Center. http://www.afdc.energy.gov/locator/stations/.

DOE/EPA (U.S. Department of Energy and U.S. Environmental Protection Agency). 2014a. "Compare Side-by-Side: 2014 Tesla Model S (85 kW-hr battery pack)." http://www.fueleconomy.gov/feg/Find.do?action=sbs&id=34775. Accessed January 26, 2015.

DOE/EPA. 2014b. "Compare Side-by-Side: 2014 Nissan Leaf." http://www.fueleconomy.gov/feg/Find.do?action=sbs&id=34699. Accessed January 26, 2015.

DOE/EPA. 2014c. "Compare Side-by-Side: 2014 Ford Focus Electric." http://www.fueleconomy.gov/feg/Find.do?action=sbs&id=34130. Accessed January 26, 2015.

DOE/EPA. 2014d. "Compare Side-by-Side: 2014 Chevrolet Volt." http://www.fueleconomy.gov/feg/Find.do?action=sbs&id=33900. Accessed January 26, 2015.

DOE/EPA. 2014e. "Compare Side-by-Side: 2014 Toyota Prius Plug-in Hybrid." http://www.fueleconomy.gov/feg/Find.do?action=sbs&id=34516. Accessed January 26, 2015.

Duoba, M. 2012. "Evaluating Plug-In Vehicles (PHEV & BEV) Using Standard Dynamometer Protocols." Presentation at the 6th U.S.-China Electric Vehicles and Battery Technology Workshop, Boston, MA, August 22-24. http://www.cse.anl.gov/us-china-workshop-2012/pdfs/session4b_demos_standards/duoba_4B-2-Duoba-ANL-Standardizing-Vehicle-Dyno-Test-Aug22-2012.pdf.

EPA (U.S. Environmental Protection Agency). 2012. "Six Common Air Pollutants." http://www.epa.gov/oar/oaqps/urbanair. Accessed November 14, 2014.

EPA. 2014. "2014 Fuel Economy Datafile." Office of Transportation and Air Quality. http://www.fueleconomy.gov/feg/download.shtml. Accessed January 26, 2015.

EPA/NHTSA (U.S. Environmental Protection Agency and National Highway Traffic Safety Administration). 2012. Joint Technical Support Document, Final Rulemaking 2017-2025 Light-Duty Greenhouse Gas Emission Standards and Corporate Average Fuel Economy Standards. Report No. EPA-420-R-12-901.

FHWA (Federal Highway Administration). 2011. Summary of Travel Trends: 2009 National Household Travel Survey. FHWA-PL-11-02. U.S. Department of Transportation. http://nhts.ornl.gov/2009/pub/stt.pdf.

Ford Motor Company. 2014. "2014 Ford Focus Electric Power and Handling." http://www.ford.com/cars/focus/trim/electric/. Accessed January 26, 2015.

Francfort, J. 2013. "U.S. Department of Energy's Vehicle Technologies Program. Plug-in Vehicles and Charging Infrastructure Usage Patterns: Lessons Learned from the First Two Years." Presentation at SAE Government/Industry Meeting, Washington, DC, January 31. http://avt.inel.gov/pdf/prog_info/SAEGovtIndustryFeb2013.pdf.

Ge, M., X. Fang, J. Rong, and C. Zhou. 2013. Review of porous silicon preparation and its application for lithium-ion battery anodes. Nanotechnology 24(42): 422001.

Goodenough, J.B., and K. Mizushima. 1981. "Electrochemical Cell with New Fast Ion Conductors." U.S. Patent 4,302,518, filed March 31, 1980, and issued November 24, 1981.

Gordon-Bloomfield, N. 2013. "Nissan Testing New Battery Pack to Address Problems with LEAF in Hot Weather." Plug-inCars,August26.http://www.plugincars.com/nissan-testing-new-battery-pack-leaf-128088.html.

Hensley, R., J. Newman, and M. Rogers. 2012. "Battery Technology Charges Ahead." McKinsey Quarterly, July. http://www.mckinsey.com/insights/energy_resources_materials/battery_technology_charges_ahead.

Hidary, J. 2012. "New Models of Mobility and EV Deployment." Presentation to the Committee on Overcoming Barriers to Electric-Vehicle Deployment, Washington, DC, December 18.

Howell, D. 2013. U.S. Battery R&D Progress and Plans. U.S. Department of Energy. http://energy.gov/sites/prod/files/2014/03/f13/es000_howell_2013_o.pdf.

Julien, C.M., A. Mauger, K. Zaghib, and H. Groult. 2014. Comparative issues of cathode materials for Li-ion batteries. Inorganics 2(1): 132-154.

Kam, K., and M. Doeff. 2012. Electrode materials for lithium ion batteries. Materials Matters 7(4). http://www.sigmaaldrich.com/technical-documents/articles/material-matters/electrode-materials-for-lithium-ion-batteries.html.

MacKenzie, A. 2013. "2013 Motor Trend Car of the Year: Tesla Model S Shocking winner: Proof Positive That America Can Still Make (Great) Things." MotorTrend, January. http://www.motortrend.com/oftheyear/car/1301_2013_motor_trend_car_of_the_year_tesla_model_s/.

Maruyama, T. 2013. "Change of Sales Price of the Battery from January 1999 to September 2012." Presentation to the Committee on Overcoming Barriers to Electric-Vehicle Deployment, Tokyo, Japan, December 9.

Masson, L. 2013. "European Commission Backs Meenekes Type 2 Electric Car Plug." Plug-in Cars, January 30. http://www.plugincars.com/european-commission-wants-act-help-evs-126265.html.

Miller, J.M, O.C. Onar, C. White, S. Campbell, C. Coomer, L. Seiber, R. Sepe, and A. Steyerl. 2014. Demonstrating dynamic wireless charging of an electric vehicle: The benefit of electrochemical capacitor smoothing. IEEE Power Electronics Magazine 1(1): 12-24.

NEC (National Electrical Code). 2008. NEC2008 National Electrical Code, Article 625. Electrical Vehicle Charging System Equipment, Section 625-2. http://www.freenec.com/T504.html.

Nelson, P.A., K.G. Gallagher, I. Bloom, and D.W. Dees. 2011. Modeling the Performance and Cost of Lithium-Ion Batteries for Electric-Drive Vehicles. No. ANL-12/55. http://www.cse.anl.gov/batpac/files/BatPaC%20ANL-12_55.pdf.

NRC (National Research Council). 2011. Assessment of Fuel Economy Technologies for Light-Duty Vehicles. Washington, DC: The National Academies Press.

NRC. 2013. Transitions to Alternative Vehicles and Fuels. Washington, DC: The National Academies Press.

NREL (National Renewable Energy Laboratory). 2013. "Plug-In Hybrid Electric Vehicles." Vehicle Systems Analysis. http://www.nrel.gov/vehiclesandfuels/vsa/plugin_hybrid.html. Accessed March 14, 2013.

ORNL (Oak Ridge National Laboratory). 2012. Wide Band Gap Devices: Powering the Next Generation of Electric Traction Drive Systems. http://web.ornl.gov/sci/ees/transportation/pdfs/WBGBroch.pdf.

Pesaran A., S. Santhanagopalan, and G. Kim. 2013. "Addressing the Impact of Temperature Extremes on Large Format Li-Ion Batteries for Vehicle Applications." Presentation at the 30th International Battery Seminar, Ft. Lauderdale, FL, March 11. http://www.nrel.gov/vehiclesandfuels/energystorage/publications.html.

Pham, H., M. Lee, K. Nam, E. Hwang, Y. Kwon, and S. Song. 2014. "High-Voltage Electrolyte Additive for High-Energy Lithium-Ion Batteries." Presentation at the 17th International Meeting on Lithium Batteries, Como, Italy, June 13. https://ecs.confex.com/ecs/imlb2014/webprogram/Paper29242.html.

PolyPlus Battery Company. 2009. "Advanced Lithium Battery Technology." http://www.polyplus.com/liair.html. Accessed October 25, 2013.

Rahman, M.A., X. Wang, and C. Wen. 2014. A review of high energy density lithium-air battery technology. Journal of Applied Electrochemistry 44(1): 5-22.

Rawson, M., and S. Kateley. 1998. "Electric Vehicle Charging Equipment Design and Health and Safety Codes." California Energy Commission. Sacramento, CA, August 31. http://www.energy.ca.gov/papers/98-09-23_KATELEY.PDF.

SAE (Society of Automotive Engineers). 2012. "SAE Electric Vehicle and Plug in Hybrid Electric Vehicle Conductive Charge Coupler." Standard Code: J1772, Revision B. SAE International, October 15. http://standards.sae.org/j1772_201210/.

Sakti, A., J. Michalek, E. Fuchs, and J. Whitacre. 2014. A techno-economic analysis and optimization of Li-ion batteries for light-duty passenger vehicle electrification. Journal of Power Sources 273: 966-980.

Schey, S. 2013. The EV Project: Q2 2013 Report. Electric Transportation Engineering Corporation. http://www.theevproject.com/cms-assets/documents/127233-901153.q2-2013-rpt.pdf.

Sion Power. 2014. "Technology Overview." http://sionpower.com/technology.html. Accessed October 25, 2014.

Sriramulu, S., and B. Barnett. 2013. "Technical, Manufacturing, and Market Issues Associated with xEV Batteries." Presentation to the Committee on Overcoming Barriers to Electric-Vehicle Deployment, Washington, DC, August 13.

Steffke, K., S. Inguva, D. Van Cleve, and J. Knockeart. 2008. "Accelerated Life Test Methodology for Li-Ion Batteries in Automotive Applications." SAE Technical Paper 2013-01-1548, doi:10.4271/2013-01-1548.

Tamor, M. 2012. "Overcoming Barriers to Deployment of Electric Vehicles NAE Committee Discussion." Presentation to the Committee on Overcoming Barriers to Electric Vehicle Deployment, Washington, DC, December 17.

Tesla. 2014a. "Increasing Energy Density Means Increasing Range." http://www.teslamotors.com/roadster/technolgy/battery. Accessed on March 30, 2014.

Tesla. 2014b. "Charge Your Model S." http://www.teslamotors.com/charging#/basics. Accessed November 14, 2014.

Tesla. 2014c. "100 Supercharger Stations." http://www.teslamotors.com/blog/100-supercharger-stations.

Toyota Motor Corporation. 2014. "2014 Prius Plug-In." http://www.toyota.com/content/ebrochure/2014/prius-plug-in_ebrochure.pdf.

Trefis Team. 2014. "Gigafactory Will Cost Tesla $5 Billion but Offers Significant Cost Reductions." Forbes, March 11. http://www.forbes.com/sites/greatspeculations/2014/03/11/gigafactory-will-cost-tesla-5-billion-but-offers-significant-cost-reductions/.

Ward, J. 2013. "DOE Electric Vehicle Activities Update." Presentation to the Committee on Overcoming Barriers to Electric-Vehicle Deployment, Washington, DC, August 13.

Woodard, T. 2012. "Overcoming Barriers to Electric-Vehicle Deployment Workshop Barriers to Deployment, an OEM Perspective." Presentation to the Committee on Overcoming Barriers to Electric-Vehicle Deployment, Washington, DC, December 17.

Yoshino, A. 2012. The birth of the lithiumion battery. Angewandte Chemie International Edition 51(24): 5798-5800.

3

Understanding the Customer Purchase and Market Development Process for Plug-in Electric Vehicles

The process of buying a vehicle is a complex, highly involved consumer decision (Solomon 2014). A vehicle is one of the most expensive purchases made by individuals or households, often equal to many months or even years of income, and will last for many years. As a result, consumers perceive the decision to be a relatively risky one and will strive to ensure a "safe" decision so that they are not stuck with a poor purchase choice for years to come. In general, consumers want vehicles that are affordable, safe, reliable, and comfortable for travel and meet many practical needs, such as getting them to work, school, stores, and recreation and vacation areas. Some also want vehicles to meet their psychosocial needs; for example, vehicles can serve as status symbols that represent one's success or self-image. For all these reasons, consumers generally will undertake lengthy research into their options to ensure a good choice that satisfies all their various needs.

Plug-in electric vehicles (PEVs) must compete effectively with internal-combustion engine (ICE) vehicles in meeting consumer needs. However, PEVs, many of which are in their first generation of deployment, add complexity and uncertainty to the consumer's multistep and potentially time-consuming process of purchasing a vehicle. Under conditions of uncertainty and perceived risk, consumers tend to gravitate to the known and familiar. That observation is well-documented in the literature, particularly in Daniel Kahneman's (2013) work, *Thinking Fast and Slow*, which spurred much recent work in behavioral economics. Because innovative products require a higher degree of learning than existing products, the effort customers must put into the decision process is greater than for more familiar products. To unseat incumbent technologies, the new technology must offer advantages and benefits sufficient to offset any price differential and the perceived risk and uncertainty of purchasing an innovation (Aggarwal et al. 1998). Thus, the committee emphasizes that consumer considerations loom large for the deployment of PEVs in the nation's transportation mix, and understanding consumer perceptions, knowledge, and behavior are key to crafting viable strategies for successful commercialization of PEVs.

This chapter begins with a general discussion of models of adoption and diffusion of innovation. It presents evidence on how new technologies are adopted by various categories of customers and discusses the factors that affect the pace of adoption and diffusion of a new technology through society. Next, the chapter discusses consumer demographics and evaluates the implications of that information and other factors that affect adoption and diffusion of PEVs. The chapter then reviews what motivates the purchases of mainstream consumers and possible barriers for their adoption of PEVs. Next, the chapter reviews strategies for addressing consumer concerns and describes government efforts to familiarize the public with PEVs. Throughout the chapter, at the conclusions of the various sections, the committee highlights relevant findings. Recommendations for addressing consumer perceptions (or misperceptions) and barriers to adoption are presented in a section dedicated to overcoming the challenges. The committee notes that the chapter focuses primarily on private (individual) new vehicle buyers, who are responsible for about 80 percent of all new vehicle purchases. Fleet sales, which average 20-22 percent of the U.S. market (Automotive Fleet 2013), are addressed at the conclusion of this chapter.

UNDERSTANDING AND PREDICTING THE ADOPTION OF NEW TECHNOLOGIES

Models for the Adoption of Innovative Products

Developers of new technologies generally, and of PEVs specifically, face challenges in developing a market and motivating consumers to purchase or use their products (Mohr et al. 2010). Incumbent technologies—in this case, ICE vehicles—can be difficult to unseat; they have years of production and design experience, which make their production costs lower than those of emerging technologies and thus more affordable. In addition, ICE vehicle technology is continuously improving; many of these improvements, which are being made to meet tighter fuel economy and greenhouse gas emission standards (EPA/NHTSA 2012), are described in the NRC report *Transitions to Alternative Vehicles and*

Fuels (NRC 2013a) . The necessary infrastructure—including dealerships, service stations, roadside assistance, and the ubiquity of over 100,000 gasoline stations across the United States (U.S. Census Bureau 2012)—is also well developed. Consumers know the attributes and features to compare to evaluate their ICE-vehicle choices, and they are accustomed to buying, driving, and fueling these vehicles. Indeed, one of the main challenges to PEV adoption is how accustomed people are to ICE vehicles.

Traditional consumer-adoption models predict the diffusion of new innovations through society (Parasuraman and Colby 2001; Rogers 2003; Moore 2014). The models are well established and empirically validated across many product categories (Sultan et al. 1990) and can help in understanding the consumer purchase decision and market development process for PEVs. As stated in Chapter 1, PEV sales reached about 0.76 percent of the U.S. market in 2014 (Cobb 2015). To put that in perspective, it took 13 years for hybrid electric vehicles (HEVs) to exceed 3 percent of annual new light-duty vehicle sales in the United States (Cobb 2013).[1]

To compare various rates of market penetration, Figure 3-1 shows the consumer technologies with the fastest growth rates. As the figure shows, new products can take many years to be adopted by a large percentage of the consumers in a market. For example, consider that the microwave—a relatively inexpensive and practical item with no complicated infrastructure needs—took 15 years to reach just 50 percent market penetration. Consumers did not have experience with microwave ovens nor did they initially see the value or usefulness of such a product; its means of cooking was not understood, and it did a poor job of "baking" compared with conventional ovens. Indeed, calling the microwave an "oven" was probably an error, as that term confused consumers about the microwave's functions. Initial uses of the microwave were to heat water, thaw and heat frozen food, and reheat leftovers—few of these tasks had much to do with how conventional ovens were used. It took many years to educate the consumer about exactly what a microwave could do. Consumer knowledge, societal lifestyle changes, and lower prices due to volume production over decades resulted in microwaves being a primary appliance in the household, nearly 20 full years after they were first introduced.

One insight is that adoption and diffusion of new innovations can be a long-term, complicated process that is especially slow for products that cost tens of thousands of dollars and where consumers have questions about infrastructure availability, resale value, and other variables. A further complication can be the innovation ecosystem, which includes all elements of the total customer solution. For PEVs, the innovation ecosystem includes not only the vehicle but also the charging stations (whether at home, at work, or in public spaces) and the necessary permitting and installation, availability of roadside assistance, and other ownership or main-

[1] More information on vehicle technologies, emissions, and fuel economy trends is available in the EPA Trends Report (EPA 2014).

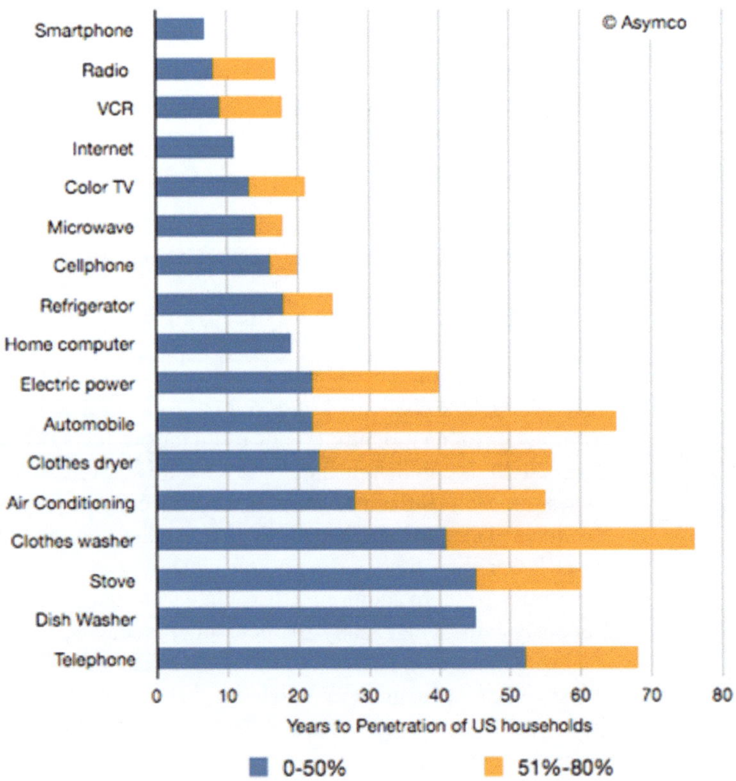

FIGURE 3-1 Years needed for fastest growing consumer technologies to achieve penetration (0-50 percent or 51-80 percent). SOURCE: Dediu (2012) © Horace Dediu, Asymco.

tenance concerns. Accordingly, the innovation ecosystem for PEVs has its own transition barriers that must be addressed for maximum market penetration to occur. Adner (2006) suggests that wide-scale deployment of new technologies is a function of three aspects of infrastructure development: (1) product technology—for example, viable, low-cost battery technology; (2) downstream infrastructure—for example, dealers, repair facilities, emergency roadside services, and battery recycling options; and (3) complementary infrastructure—for example, charging stations (whether residential, workplace, or public), knowledgeable electricians, and amenable zoning and permitting at the municipal level.

Adner's work on innovation ecosystems provides guidance for how industry stakeholders might make investment decisions to encourage adoption of new technologies. For example, if infrastructure is identified as the critical bottleneck that affects customer adoption and use, industry stakeholders might decide to invest more in infrastructure development than in the product itself. Indeed, Japan has recognized that need and has instituted a major initiative to build an extensive charging infrastructure to instill range confidence and ensure a safety net for limited-range battery electric vehicle (BEV) drivers (METI 2010). Brown et al. (2010) also emphasized the importance of supporting infrastructure development and advocated for standardization of codes, training, and other aspects of infrastructure to facilitate the PEV market.

Given the complexity of the innovation ecosystem, mainstream consumers typically are unwilling to undertake what might be perceived as a risky purchase until all elements of the requisite infrastructure are in place (Moore 2014). Indeed, if all aspects of the innovation ecosystem are not ready when consumers are making purchase decisions, industry adoption rates can be substantially lower than initial expectations.

Adoption and diffusion models provide insight into what might be considered realistic expectations about market penetration rates. Given that about 16 million new vehicles are purchased each year, it would take at least 16 years to convert the total U.S. fleet of 250 million passenger vehicles and light-duty trucks if only PEVs were sold. In addition, not all households exhibit the demographic and lifestyle traits that make PEVs a viable purchase option. Specifically, when estimating the total addressable market for PEV sales, one must consider what percentage of the total population would find PEVs practical for their travel patterns and needs. A nationally representative telephone survey of adult vehicle owners found that 42 percent of drivers—45 million households—meet the basic criteria[2] necessary to use a plug-in hybrid electric vehicle (PHEV), such as the Chevrolet Volt, for their transportation needs with few, if any, changes in behavior (Consumers Union and the Union of Concerned Scientists 2013). Of the drivers who could use a PHEV, 60 percent also fit the profile of those who could use a limited-range BEV, such as the Nissan Leaf, without major life changes.

Therefore, market adoption and diffusion of PEVs, which are expensive, infrequently purchased, long-lasting products with a complicated industry ecosystem, will be a slow process that will take decades. That insight is corroborated by the data presented early in Chapter 1 regarding early-market growth rates and market shares of PEVs.

Finding: Market penetration for new technology—particularly expensive, infrequently purchased, long-lasting innovations with a complicated ecosystem—is typically a slow process that takes 10-15 years or more to achieve even nominal penetration.

Finding: Market penetration rates are a function not only of the product being purchased but also of the entire industry ecosystem. Hence, product technologies, downstream infrastructure, and complementary infrastructure all must be attended to simultaneously during the development process.

Finding: PEVs on the market as of 2014 are not a viable option for all vehicle owners; rather, perhaps only about 40 percent of U.S. households exhibit lifestyles amenable to owning and operating a PEV.

Consumer Diffusion Models and Market Segments

Diffusion models categorize consumers (adopters) on the basis of their propensity to adopt new technologies and identify the factors that facilitate adoption and diffusion. Figure 3-2 illustrates that markets for innovation comprise five distinct categories of adopters; Table 3-1 describes each category in terms of demographic and psychographic characteristics and buying motivations. Psychographics refer to values and lifestyles of consumers and can be determined empirically through market research on their activities, attitudes, interests, and opinions (Kahle and Chiagorous 1997; Wells 2011). Although demographics can explain *who* is buying particular types of products, psychographics are more likely to explain *why* customers buy; therefore, psychographics generally are more useful than demographics in understanding customer decisions. Major factors that affect diffusion include communication (word-of-mouth) between consumers and social networks (Mahajan et al. 1990).

DEMOGRAPHICS AND IMPLICATIONS FOR ADOPTION AND DIFFUSION OF VEHICLES

Demographic Traits of Buyers of Plug-in Electric Vehicles

Demographic traits of PEV buyers are compared with those of ICE-vehicle buyers in Table 3-2, which shows that many characteristics of PEV buyers correspond to the traits

[2] Basic criteria for PHEVs independent of pricing included access to parking and an electric outlet at home or work, seating capacity for no more than four occupants, and no hauling or towing capability. A BEV was considered suitable not only when the PHEV criteria were met but also when the maximum weekday driving distance was less than 60 miles and other household vehicles were available if weekend driving frequently exceeded the current BEV range (Consumers Union and the Union of Concerned Scientists 2013).

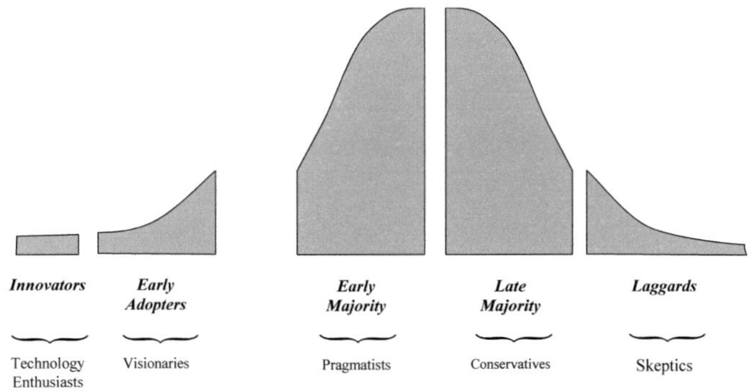

FIGURE 3-2 Distribution of adopter categories. Labels reflect orientation to technology. SOURCE: Moore (2014) ©1991 by Geoffrey A. Moore. Reprinted by permission of HarperCollins Publishers.

TABLE 3-1 Categories and Descriptions of Adopters

Category	Description
Innovators or enthusiasts	Are technology enthusiasts or lovers. Are willing to buy early release versions even if product quality or reliability are not yet proven or established. Want to work with developers and infrastructure providers to improve new products, a source of pride in their own techno-intelligence. Are important segments for endorsement about viability of the new innovation category. Are not a large enough market segment to be a long-lived or significant source of revenue.
Early Adopters or Visionaries	Are less concerned about price and more motivated by psychosocial benefits, such as visibility of their purchase in their peer group. Are more affluent, cosmopolitan, and, typically, younger than other categories. Are willing and motivated to address early market development problems, including service and infrastructure challenges, which when solved, become a source of pride. Are generally considering or comparing purchases not within the product category (for example, with a different vehicle make or model) but with some other major purchase.
Early Majority or Pragmatists	Are very concerned about value (benefits received relative to price paid). Want to evaluate several different models or options within the product category. Are willing to purchase only when all elements of the requisite infrastructure are in place. Want a hassle-free solution that performs as promised. Are not willing to tolerate anxiety or doubt. Are first sizable segment of the market by volume.
Late Majority or Conservatives	Tend to buy when there are a plethora of models and choices in the market and when prices have substantially decreased.
Laggards or skeptics	Would prefer not to buy anything designated as a new technology. Do so only when they can no longer avoid doing so.

NOTE: *Early* and *late* are relative terms based on the time it takes to adopt.

of early market adopters. PEV buyers had a median income of nearly $128,000 to $148,000 whereas ICE-vehicle buyers had a median income of about $83,000 (Strategic Vision 2014). By way of comparison, HEV buyers had a median household income of $90,204, and the average median U.S. household income was $51,017 (U.S. Census Bureau 2013a). Consistent with traits of early adopters, PEV buyers were better educated than ICE-vehicle buyers.

Table 3-3 lists demographic data for purchasers of a vehicle from each category of PEV as defined in Chapter 2 (long-range BEV, limited-range BEV, range-extended PHEV, and minimal PHEV) compared with data for all new-vehicle buyers. The table shows that of the four types of PEVs, Tesla Model S buyers are primarily men who have higher incomes, paid cash, and did not seriously consider purchasing another vehicle, whereas Nissan Leaf buyers are younger with larger household sizes. Chevrolet Volt buyers exhibit lower educational levels than other PEV buyers. Toyota Plug-in Prius buyers have a higher percent of female buyers. Finally, PEV buyers who considered other models of PEVs in their purchase process reported that the vehicle that they most seriously considered was the Chevrolet Volt.

The data presented in Table 3-3 also show that leasing rates vary by PEV model. The Nissan Leaf and Chevrolet Volt have higher lease rates than the Toyota Plug-in Prius or Tesla Model S (Strategic Vision 2014). Furthermore, data from the California Plug-in Electric Vehicle Survey indicate that PEVs are leased at a rate of 28.8 percent in California,

TABLE 3-2 Comparison of New BEV Buyers, PHEV Buyers, and ICE-Vehicle Buyers

Characteristic	BEV Buyer	PHEV Buyer	ICE-Vehicle Buyer
Gender	77% male	70% male	60% male
Marital status	81% married	78% married	66% married
Average age	48 years	52 years	52 years
Education	86% college graduate	77% college graduate	59% college graduate
Occupation	42% professional	37% professional	25% professional
Median household income	$148,158	$127,696	$83,166
Number of respondents	3,556	1,000	186,662

NOTE: BEV, battery electric vehicle; ICE, internal-combustion engine; PHEV, plug-in hybrid electric vehicle.
SOURCE: Strategic Vision New Vehicle Experience Study of Vehicle Registrants, October 2013-June 2014.

TABLE 3-3 Comparison of All New-Vehicle Buyers to Buyers of Specific Plug-in Electric Vehicles[a]

Characteristic	All New-Vehicle Buyers	Tesla Model S	Nissan Leaf	Chevrolet Volt	Toyota Prius Plug-in
Gender (M/F)	61/39	82/18	77/23	74/26	66/34
Married or partnered	71	83	87	82	76
Age 50+	56	68	37	61	39
Household size of 1 or 2	58	56	35	53	46
College grad or more	59	87	86	77	83
Income +$100K	40	88	66	63	62
Caucasian	79	86	70	82	56
Purchased/leased	78/22	95/5	14/86	56/44	68/32
Paid cash	14	36	5	12	2
Received special financial incentives	64	24	76	73	88
Did not seriously consider any other vehicle	NA	62	50	42	48
Seriously considered other models	NA	Chevrolet Volt (1%)	Chevrolet Volt (10%)	Toyota Plug-in Prius (5%)	Chevrolet Volt (8%)
Number of respondents	237,235	285	2,257	556	169

[a] Entries are provided as percent of respondents.
SOURCE: Strategic Vision New Vehicle Experience Study of Vehicle Registrants, October 2013-June 2014.

greater than the overall lease rate for light-duty vehicles in the United States (Rai and Nath 2014; Tal et al. 2013). Although many consumers have never leased a vehicle and are therefore unfamiliar with the process, leasing a PEV removes the risk to the consumer that is associated with unknown resale value, battery decay, and rapid technology changes. Moreover, leasing agencies are able to incorporate the federal tax incentives into a shorter period of time. As a result, attractive leasing deals have positively affected PEV sales (Loveday 2013a). Whether leases appeal differentially to early adopters or mainstream customers is unknown.

To date, male buyers dominate the PEV market. Figure 3-3 shows that although women make between 50 and 60 percent of vehicle purchases generally (the top two bars in the figure represent U.S. data on all vehicles), their involvement in PEV purchases ranges between only 15 and 30 percent (the bottom four bars in the figure represent data on California PEV buyers or lessees only) (Caperello et al. 2014). The authors' detailed interviews and focus groups find that men treat PEV purchases as "projects"—a classic feature of early market adopters—whereas women in the study expressed more practical concerns and did not want to experiment, a buying trait more typical of mainstream adopters. Hence, the gender data also are consistent with differences between early adopters and mainstream adopters.

Finding: PEVs to date have been sold primarily to customers in the early adopter segment of the marketplace whose

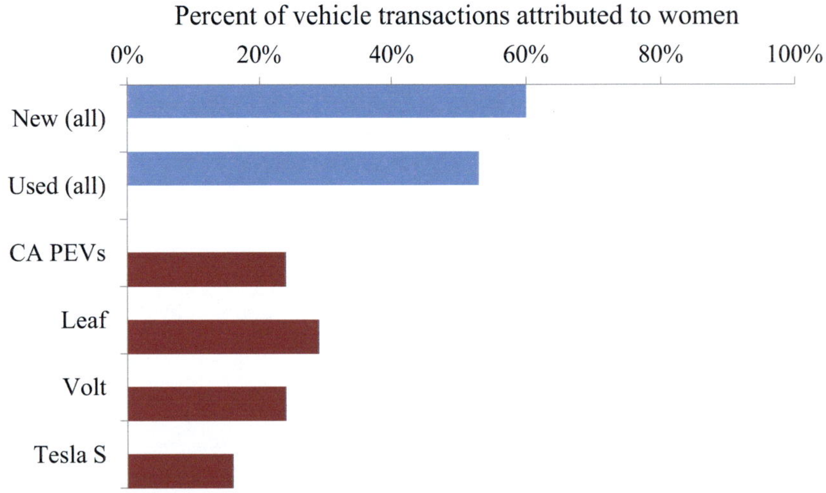

FIGURE 3-3 Women's rate of participation in the markets for all vehicles and for PEVs. The figure shows that women's participation in vehicle purchases is much lower for PEVs than for vehicles as a whole. Data in blue represent the entire used and new-vehicle market for the entire United States. Data in red reflect California PEV purchasers. NOTE: PEV, plug-in electric vehicle. SOURCE: Image courtesy of Kenneth S. Kurani, University of California, Davis, Institute of Transportation. Data compiled from NBCUniversal, Center for Sustainable Energy, California Air Resources Board Clean Vehicle Rebate Project, and EV Consumer Survey Dashboard.

traits and buying motives are different from those of the mainstream market segment.

Selecting a Beachhead

Diffusion is a social and geographic process; at any point in time, diffusion in one region of a large country can be ahead of diffusion in another, as is illustrated in Figure 3-4, which shows the variation in PEV deployment across the United States and provides the projected cumulative PEV volume in 2014 for the 100 largest urban areas. PEVs tend to be sold in states and municipalities where both the demographic and psychographic profiles of residents are consistent with those of the early adopter category; these areas also tend to have a positive regulatory climate for PEVs. California is one such area and has a long history of strong sales for new vehicle technologies. It has the highest proportion of HEVs in the United States, and the Toyota Prius hybrid was the best-selling vehicle in California in 2012 and 2013.

To ensure that new technologies succeed with mainstream consumers, Moore (2014) suggests selecting a "beachhead," a narrow market segment of consumers for whom the new technology offers "a compelling reason to buy." That approach is in contrast to conventional thinking that a broad mass market is desirable. The logic behind a beachhead is that, by offering a compelling value proposition specifically targeted to meet the needs of a narrow subset of consumers, the technology stands a greater chance of dominance in a key market segment. Then, the momentum gained through dominance in the initial beachhead can be used more efficiently and effectively to drive sales in related, adjacent segments. For example, word-of-mouth communication is easier and more effective between adjacent market segments (related geographically, by common lifestyles, or by common professional circles) because people will find communicating with others who have similar traits more credible and relevant than with those who have dissimilar traits. Thus, rather than attempting to succeed in the broad mass market, providers of new and complex technologies find it advantageous to focus on a narrow segment of consumers for whom the innovation offers a compelling reason to buy. Success in that initial segment then can be leveraged powerfully in adjacent segments.

For the PEV market, a beachhead approach logically would focus on key geographic regions or regional corridors where momentum has already been established; infrastructure is more readily available; word-of-mouth between neighbors, friends, and co-workers can occur more readily; where there is greater availability of PEV makes and models; and where gasoline is expensive or electricity is cheap. As one might expect, California is a particularly attractive market; it accounts for over one-third of annual PEV sales in the United States, and sales of PEVs in California at the close of 2014 comprised 3.2 percent of new light-duty vehicle sales and 5.2 percent of new passenger vehicles (CNCDA 2015). It also has a supportive regulatory environment with its zero-emission-vehicle (ZEV) mandate, which has been a prime contributor to the availability of PEV models in California. States that have agreed to implement the multistate ZEV

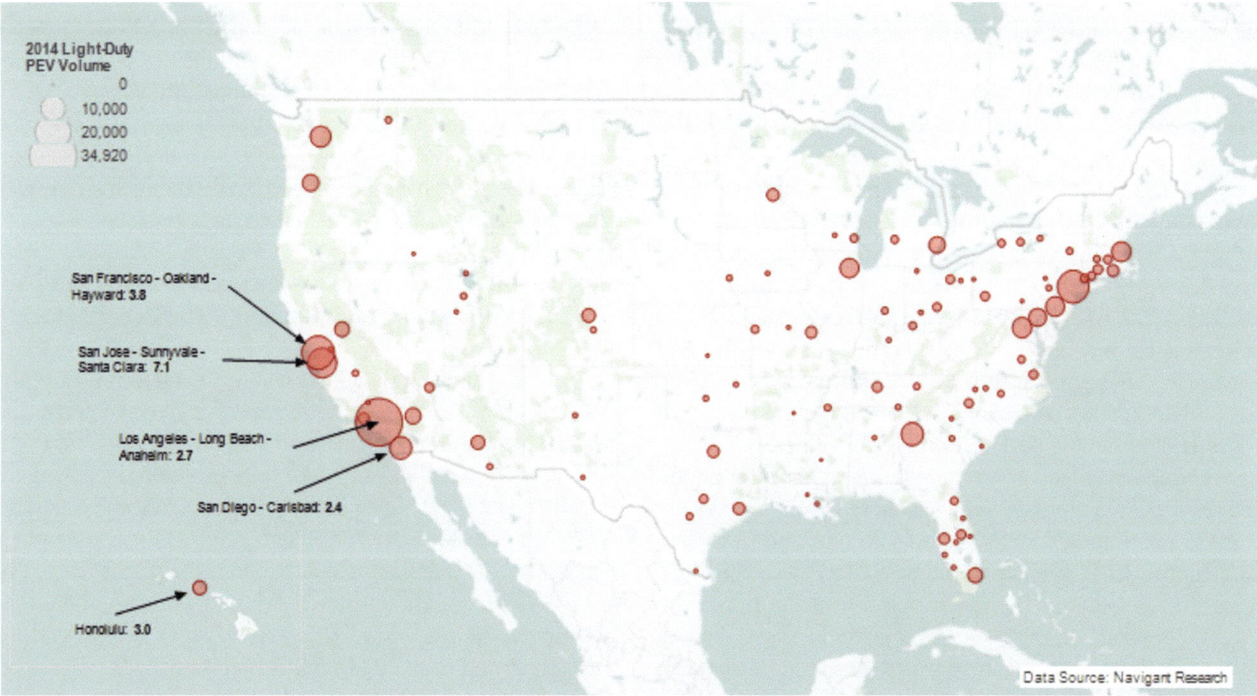

FIGURE 3-4 Projected 2014 light-duty PEV volume in the 100 largest MSAs. Dot size is proportional to the projected total PEV volume in that MSA in 2014. It can be seen that large numbers of vehicles are located in Los Angeles, the San Francisco Bay Area, New York, San Diego, Seattle, and Atlanta. As noted in the figure, San Jose, San Francisco, Honolulu, Los Angeles, and San Diego have the largest per capita concentrations of PEVs (volume projected for 2014 per 1,000 people based on 2012 census projections). NOTE: MSA, metropolitan statistical areas; PEV, plug-in electric vehicle. SOURCE: Data courtesy of Navigant Research in Shepard and Gartner (2014).

action plan (modeled after California's ZEV mandate) and that have greater availability of PEV models are also favorable places for the beachhead approach. They include Connecticut, Maryland, Massachusetts, New Jersey, New York, Oregon, and Rhode Island.[3] Places where there is clean and low-cost hydroelectric power are also favorable locales for the beachhead approach. One such example is Washington state, which has higher PEV per capita sales than California.

One final segment that could constitute a favorable PEV market is the multiple-vehicle household. Most households have more than one vehicle. At the national level, of the roughly 75 million owner-occupied housing units, 3.4 percent have no vehicle, 26.7 percent have one vehicle, 43.8 percent have two vehicles, and 26.1 percent have three or more vehicles (U.S. Census Bureau 2013b). Having multiple vehicles offers the opportunity to choose among vehicles that have different utilities. For example, a multiple-vehicle household might be able to mitigate the challenges of owning a limited-range BEV if it also owns a PHEV or an ICE vehicle that can be used for long-distance trips. Many households that have multiple vehicles, however, might not be able to replace all their vehicles with PEVs because all the parking spots might not have access to charging infrastructure.

Finding: The PEV market is characterized by strong regional patterns that reflect certain key demographics, values, and lifestyle preferences and have favorable regulatory environments for PEVs.

Finding: Initial beachheads for PEV deployment are specific geographic areas, such as California, that have expensive gasoline; key demographics, values, and lifestyles; a regulatory environment favorable to PEVs; a variety of PEV makes and models available; and existing infrastructure or an ability to readily deploy such infrastructure.

Driving Characteristics and Needs of the Mainstream Consumer

As discussed, selecting a beachhead plants the seeds for the diffusion and adoption of PEVs, but PEVs will need to meet more consumer needs to gain greater market share or become widely adopted. To understand what mainstream

[3] Information on model availability by state was provided by representatives of vehicle manufacturers. Sources were Brian Brockman, Nissan, September 8, 2014; William Chernicoff, Toyota, August 22, 2014; Kevin Kelly, Joe LaMuraglia, and Shad Blanch, GM, August 22, 2014; James Kliesch, Honda, September 2, 2014; Nancy Homeister, Ford, September 2, 2014; and Dan Irvin, Mitsubishi, September 10, 2014.

consumers might want, it is important to consider how people use vehicles and how those driving habits intersect with the four classes of PEVs defined in Chapter 2. As of 2013, there were more than 233 million light-duty vehicles registered in the United States, each traveling on average 11,346 miles per year (FHWA 2015). The Federal Highway Administration provides more detailed information about household trips that might help to determine whether consumers would be interested in purchasing and using PEVs. In the most recent data from 2009, households reported an average of 3.02 trips per vehicle per day and 28.97 miles per day per vehicle and an average vehicle trip length of 9.72 miles (FHWA 2011). Changing trends in vehicle ownership and use are discussed in greater detail in the next section.

Although averages provide some important information about how people use their vehicles, there is substantial variability in use among drivers and vehicle type and over time (for example, from one day to the next), so that average use might not fully capture consumer needs over the life of the vehicle. For example, the National Household Travel Survey shows that trips of fewer than 10 miles constituted 71 percent of trips and accounted for 25 percent of miles traveled. Commuting is a common routine trip that averages lengths of 6 miles and represents 27.8 percent of miles. Routine trips are important to consider because they represent an opportunity to electrify miles and maximize the value proposition for PEVs. Long trips (over 100 miles) represented less than 1 percent of trips but 16 percent of miles traveled (FHWA 2011). As noted in Chapter 2, long trips are an issue for BEVs because trips that exceed the all-electric range become inconvenient.

Mainstream consumers consider what kinds of trips they need to complete when purchasing (and using) a vehicle. Those considerations will affect their views on the utility of the vehicle. Many consumers might not find the utility of a long-range BEV to be substantially limited by trip distance. Some consumers might find that although a limited-range BEV might meet their average travel needs, it does not meet their needs to make the occasional long trip. A high frequency of those "inconvenient days" might greatly dissuade a consumer from purchasing a limited-range BEV. However, if consumers have multiple options for making longer trips, such as public transportation or a second vehicle, they might find that a limited-range BEV best meets their routine needs. PHEVs can accommodate all possible trip lengths with easy refueling, but they sacrifice electric miles for gasoline-fueled miles on longer trips. Average or routine travel needs, such as a commute, might also affect the PHEV range that a consumer might choose because matching PHEV range to average or routine use might improve the consumer value proposition. This discussion assumes that consumers understand their needs and the ability of various types of vehicles to meet those needs. Later, this chapter discusses misconceptions and gaps in knowledge about PEVs that lead to consumer misperceptions of range and vehicle utility, a barrier to PEV deployment.

Finding: Although there is substantial variability in vehicle use, average daily travel or other routine use provides a metric that can help evaluate the utility of a PEV.

Finding: Aside from average or routine use, many consumers make a small number of long-distance trips that might weigh heavily in their vehicle purchase decision.

Changing Landscape of Vehicle Ownership and Use: Implications for Adoption and Diffusion of Vehicles

Social and demographic changes are affecting the amount that people drive and the demand for new vehicles; these changes have implications for PEV sales and their use. Vehicle miles traveled (VMT) per capita generally increased from 1960 to 2007, outpacing growth in gross domestic product per capita. After 2007, VMT peaked, and VMT per capita began to decline as unemployment and gasoline prices rose, resulting in fewer commuters, fewer driving vacations, and more attention to the cost of fuel (FHWA 2012a; U.S. Census Bureau 2013a; Zmud et al. 2014). A 2014 report from the Transportation Research Board identified an aging and more ethnically diverse population, along with changing patterns in the workforce, urban living, household formation, views on environmentalism, and use of digital technology, which are affecting total and per capita VMT (Zmud et al. 2014). Although VMT is on the rise again, it is still below 2007 levels (FHWA 2014) and is projected to grow at an average annual rate of only 0.9 percent between 2012 and 2040 (EIA 2014a).

The demand for new vehicles also appears to be changing. First, the demand for new vehicles has decreased. Americans buy new vehicles every 6-8 years on average, as compared with every 3-4 years before the recession (LeBeau 2012). Related research from J.D. Power (Henry 2012) shows that the average trade-in vehicle at dealerships is now 6.5 years old, 1 year older than the average in 2007. In contrast to almost all products, vehicles have a robust secondary (used) market that is larger than the new market; in fact, two-thirds of all U.S. vehicle purchases are for used vehicles (35.7 million in 2013) (Edmunds 2013). Those data have implications for vehicle purchases generally and PEV purchases specifically. Given the length of time between purchases, product options will have changed substantially, particularly because of model and technology changes, and what the consumer might want or need in a new model might have changed substantially. Thus, the consumer likely will undertake a lengthy and exhaustive process before purchase to research new options on the market; that research could take as long as several weeks or months and involve many hours of online research before even visiting dealerships (Darvish 2013). The decreased demand for new vehicles and the lengthy research process will certainly affect the adoption and diffusion rates for PEVs.

Second, the number of households without a vehicle has increased nearly every year since 2005; it was 8.87 percent in 2005 and 9.22 percent in 2012 (U.S. Census Bureau 2013a; Sivak 2014). Fewer vehicle-owning households might mean fewer households in the market for PEVs. The percentage of households without a vehicle also varies widely by geographical area. New York City, Washington D.C., Boston, Philadelphia, San Francisco, Baltimore, Chicago, and Detroit all have more than 25 percent of households without a vehicle (Sivak 2014). The geographic variation will affect where PEVs sell well.

Another factor that is changing is household formation. In 2000, 68.1 percent of households were defined as "family" (married couples with children, married couples without children, single parents with children, or other family). In 2010, that estimate decreased to 66.4 percent because single-person households increased from 25.8 percent to 26.7 percent (U.S. Census Bureau 2013a). As single-person households and urbanization increase, charging vehicles at home could become even more complicated as people move into apartments and multifamily dwellings and away from single-family homes that have garages or dedicated parking. As urbanization continues to rise, people might use transportation modes other than the traditional ICE vehicles. Although urban dwellers tend to log fewer VMTs (a characteristic favorable to PEV ownership), they might face challenges in finding a reliable and regular place to plug in and recharge in a city.

The societal changes noted are influencing the growth of alternative transportation methods, such as on-demand transport services (for example, Uber and Lyft) and car-sharing programs (for example, Car2Go and Zipcar). According to Susan Shaheen and Adam Cohen (2013), there are about 850,000 car-sharing members and 15,000 vehicles in North America (see Figure 3-5). Frost and Sullivan estimate in their optimistic scenario that up to 7 million members and 155,000 vehicles could be part of car sharing by 2020 (Brook 2014). Given that car sharing and on-demand services are growing and tailoring their services to city living, urban consumers are becoming reluctant to assume the responsibility and expense of a vehicle.

Car sharing could be a win or a loss for PEVs, depending on whether the programs use PEVs. If they do, they would provide ways for drivers outside the new-vehicle market to use PEVs. However, if potential PEV buyers chose to use car sharing and car-sharing programs use only ICE vehicles, PEV sales could be hurt and fewer miles electrified. Car-sharing programs are discussed further later in this chapter.

Finding: Demographics, values, and lifestyles affect not only vehicle preferences but also the practicality of a given PEV for a given individual. Different market solutions will be needed for different market categories and segments.

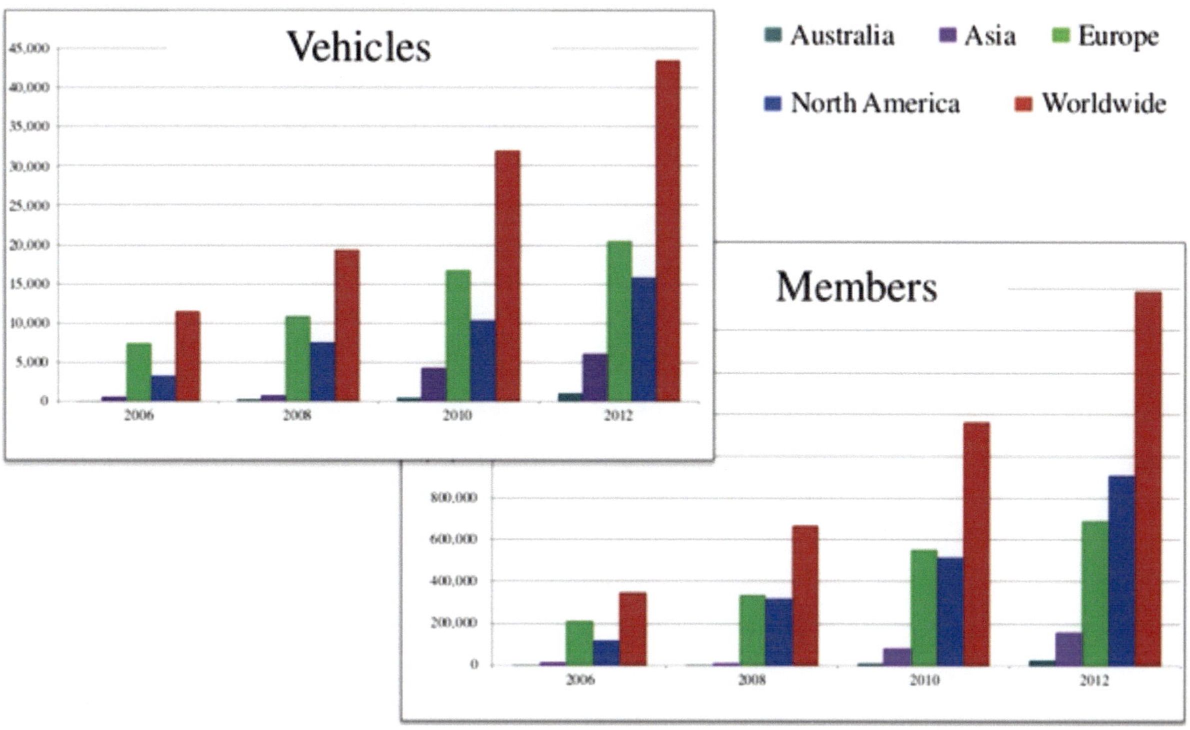

FIGURE 3-5 Worldwide growth of car sharing in terms of vehicles and members. SOURCE: Shaheen and Cohen (2013), Transportation Sustainability Research Center, University of California at Berkeley.

TABLE 3-4 Factors That Affect Adoption and Diffusion of Innovation

Factor	Description
Relative advantage	The buyer's perceived benefits of adoption (such as fuel savings) relative to the price paid (PEVs are expensive relative to ICE vehicles) and the nonmonetary costs (such as concerns about battery life, charging infrastructure, resale value, and vehicle range if a limited-range BEV).
Complexity	Difficulty of using the new product. For example, what is involved in charging at home, at work, or at public stations? Are permits required for at-home installation? Is membership needed for a charging network? How much will the electricity to fuel the vehicle cost, and how is that cost calculated?
Compatibility	How well does the new technology fit into the buyer's lifestyle? For example, is the range of a limited-range BEV adequate? Consumer concerns about standards, for example, different plug types and charging networks with different communications protocols and payment methods; mainstream consumers take a wait-and-see attitude to avoid purchasing the wrong product that does not become the dominant design.
Trial-ability	How easy is it for a potential customer to try the new technology? A typical test-drive for a PEV can demonstrate its acceleration speed and drivability but does not allow the buyer to experience charging or to resolve other concerns that inhibit purchase of a PEV.
Observability	How observable are the benefits of the new purchase to the consumer, such as fuel savings relative to electricity costs; convenience of charging at home and not having to go to a gasoline station; and quiet driving experience. How observable are the new technology and its benefits to other consumers; for example, seeing neighbors or co-workers drive a PEV or seeing PEVs plugged in at a public location hastens diffusion, much like iconic white ear buds and wires were highly visible symbols of Apple products.

NOTE: BEV, battery electric vehicle; ICE, internal-combustion engine; PEV, plug-in electric vehicle.
SOURCE: Adapted from Mohr et al. (2010).

THE MAINSTREAM CONSUMER AND POSSIBLE BARRIERS TO THEIR ADOPTION OF PLUG-IN ELECTRIC VEHICLES

Insights into strategies to diffuse new vehicle technologies beyond early adopters can be gleaned from industry studies on what consumers consider when they make a purchase and by examining general factors that affect adoption and diffusion of new technologies (Rogers 2003). Five factors typically affect the rates of adoption and diffusion for innovative products; these factors are shown in Table 3-4, which also provides implications specific to PEV deployment.

As noted earlier, the characteristics and buying motivations differ between categories of consumers. The characteristics of PEV owners to date are consistent with those of the early adopters. Because mainstream adopters (early and late majority categories combined) comprise the bulk of the purchases for any new technology (Rogers 2003), understanding their purchase motivations is critically important to increasing PEV deployment.

The top five reasons consumers give for their vehicle purchase choices generally (not specific to PEVs) are reliability, durability, quality of workmanship, value for the money, and manufacturer's reputation (Strategic Vision 2013). Although often assumed to be a key influential factor in vehicle purchases, fuel economy is a primary consideration for 45 percent of consumers (compared with reliability, a primary consideration for 68 percent of consumers). In fact, fuel economy ranked 11 of 54 reasons on the basis of Strategic Vision's May 2013 survey results. Interestingly, the average gasoline price per gallon was $4.02 at the time of the survey, and yet consumers still ranked such features as seating comfort above fuel economy as a purchase reason. Just 5 percent of U.S. consumers who purchased a vehicle responded that they were willing to pay more for an environmentally friendly vehicle (Strategic Vision 2013). Additional survey data from "rejecters" (people who considered buying a PEV but chose not to buy one) reveal consumer concerns about the reliability of the technology and the durability of the battery (Strategic Vision 2013). Those data suggest that consumers appear to use traditional criteria (reliability and durability) in their PEV evaluations and that PEVs today must compare effectively with ICE vehicles on traditional criteria to be competitive.

Egbue and Long (2012) conducted a study to explore concerns about PEVs specifically. Interestingly, in their sample of respondents (a population of faculty, staff, and students from a technically oriented university), battery range was the biggest concern expressed about PEVs, followed by the cost differential of PEVs compared with ICE vehicles. Battery range is *not* a question that is asked in typical vehicle industry research studies.[4] Corroborating the results of

[4] Business experts note several caveats in conducting and interpreting consumer research on new technologies (Leonard-Barton et al. 1995; Rayport and Leonard-Barton 1997; Seybold 2001; McQuarrie 2008). First, consumers necessarily are constrained in their responses by their knowledge of and familiarity with a given technology. Although they provide answers to research questions, the validity of their responses can be suspect. Moreover, the nature of the research protocols is similarly constrained by the known

the Strategic Vision study above, Egbue and Long (2012) similarly found that environmental considerations carry less weight in the purchase decision for PEVs than battery range and cost. Their study further suggests that even with incentives to subsidize the cost of PEVs, penetration rates are likely to remain low if consumers have low confidence in the technology.

Despite the fact that 55 percent of people shopping for a vehicle have "favorable" or "very favorable" impressions of PEVs (versus 62 percent in 2009) (Pike Research 2012), the purchase rates are still low.[5] Importantly, consumers make decisions on the basis of their perceptions rather than factual data. Astute marketers realize that consumer perceptions form the basis of their reality—even if their perceptions are factually inaccurate. Although objectively, PEVs might exhibit a lower total cost of ownership than ICE vehicles, whether consumers actually compute a total cost of ownership in making vehicle purchase decisions is not apparent.[6] Ingram (2013) states that 75 percent of people in 21 of the largest cities in the United States were unaware of cost savings and reductions in maintenance costs of PEVs. In fact, even for high-involvement purchase decisions, in which the assumption of a "rational consumer" is often made, psychosocial factors can be more important than rational considerations.

In addition to the price differential between PEVs and conventional vehicles and the range concerns for limited-range BEVs, the committee identified several additional barriers to PEV purchases—most of which are highly interrelated—that affect consumer perceptions and their decision process and ultimately (negatively) their purchase decisions. They include the limited variety and availability of PEVs; misunderstandings concerning range of PEVs; difficulties in understanding electricity consumption, calculating fuel costs, and determining charging infrastructure needs; complexities of installing home charging; difficulties in determining the "greenness" of the vehicle; lack of information on incentives; and lack of knowledge of unique PEV benefits. Those barriers are discussed briefly in the following sections.

and familiar. When replicating vehicle surveys to assess PEVs, the surveys do not include questions to assess consumer knowledge of and preference for charging infrastructure, range, and other relevant factors. As a result of those and other limitations, innovation experts recommend alternative methods of market research to complement traditional surveys and focus groups. 3M, Intel, HP, and other companies known for their culture of innovation rely on a variety of alternative research protocols, many of them more observational in nature, to recognize such limitations.

[5] As a point of reference, the same survey showed 61 percent of consumers have "a favorable or very favorable" impression of HEVs that have sales of about 3-3.5 percent of new passenger vehicle sales.

[6] Despite the lack of information, Eppstein et al. (2011) found in a simulation model that making available estimates of lifetime fuel costs associated with different vehicle types could enhance market penetration substantially. That possibility is supported by marketing in Japan, where at least one PEV manufacturer is actively using marketing messaging with information on total cost of ownership.

Limited Variety and Availability of Plug-in Electric Vehicles

Consumers are accustomed to a dizzying array of ICE vehicle models and styles available from more than a dozen manufacturers. They include performance sports cars, mid-sized passenger cars, sport utility vehicles, crossovers, luxury sedans, compact and subcompact economy cars, sporty compacts, pickup trucks, minivans, and full-sized vans. Because consumers have a wide variety of needs and motivations, a wide array of PEV makes and models are needed to satisfy them. The rather limited choice of PEVs could slow market development.

Further complicating the rather limited variety of PEVs on the market is the fact that not all PEVs are available for sale in all states. Two main considerations affect vehicle availability. One is the availability of PEVs to the dealers, which is dictated by the vehicle manufacturers. Given the questionable profit margins (Lutz 2012; Voelcker 2013a; Loveday 2013b), some vehicle manufacturers might not be motivated to offer PEVs for sale in all 50 states. The other consideration is the availability of PEVs to customers—specifically, the number of dealers in a given area actually stocking the vehicle and the number of vehicles on the lot. PEV availability is highly variable by dealer and by location. Lack of availability and the limited diversity of PEV options are barriers to consumer adoption.

Range of Plug-in Electric Vehicles

Range anxiety refers to the fear of running out of charge and being stranded. The driver's experience of range anxiety can be mild or strong and depends on the vehicle range, charging routines, and driving patterns (Frank et al. 2011). As discussed in Chapter 2, range limitation should be an issue only for limited-range BEVs. Yet, data collected from people who considered a PEV but did not buy one (rejecter data) reveal inaccurate perceptions about PEV range. For example, some buyers who considered the Chevrolet Volt did not buy it because it "lacked range," despite the fact that the Volt's onboard ICE gives the vehicle a range similar to that of a conventional vehicle. Specifically, after its 38 miles of all-electric range are depleted, it offers another 344 miles on gasoline. Such observations show that a lack of familiarity with PEVs poses a barrier to vehicle deployment; this negative effect is corroborated by the modeling work of Lim et al. (2014), who found that range concerns, as well as concerns over unknown resale value, inhibit mass adoption of PEVs.

Understanding Electricity Consumption

Drivers of ICE vehicles are accustomed to fueling with gasoline and understand how much range they have left and where gasoline stations are located relative to that range. PEV drivers, however, face a new experience—fueling with electricity—and will need to understand the interaction between several factors, including the storage capacity of the batteries, access to charging infrastructure, and driving behavior. The amount of stored electricity is measured and then communicated through dashboard displays that provide an estimate of the remaining range of the battery, a measurement that not only is new but that can also be imprecise. PEV owners will experience consuming the electric energy (depleting the battery) quickly or slowly, depending on driving speed (fast or slow), conditions (such as ambient air temperature and steepness of the road grade), and driving style of the driver (light-footed or heavy-footed) (Turrentine et al. 2011). A PEV on a cold day can consume its stored electric energy quickly because some portion of that energy goes to heat the vehicle interior; hence, drivers might see the battery energy on the dashboard display drop rapidly. For example, the range of a Nissan Leaf is 84 miles on the EPA test cycle, but if the owner drives 90 percent of his or her miles at speeds above 70 mph and lives in a cold climate, the range could be as low as 50 miles. Thus, to feel comfortable purchasing a PEV, consumers generally must understand PEV fuel consumption.

Calculating Fuel Costs

Determining electricity costs relative to gasoline costs is yet another factor that affects consumer perceptions and purchase decisions.[7] Box 3-1 shows how electricity cost could be calculated. The committee was not able to find data on consumer perceptions of electricity costs compared with gasoline costs. However, the calculations in Box 3-1 are likely complex enough to be overwhelming for a typical mainstream consumer and highlight the difficulty that consumers face in computing fuel costs, particularly compared with those for ICE vehicles. In fact, few consumers are likely to go into this level of detail to understand fuel costs when considering a vehicle purchase. The unknown costs represent yet another source of doubt and are therefore another barrier.

Overall, the data indicate that energy costs for PEVs are likely to be lower, even one-half of gasoline costs. Enrolling in special rate plans, taking advantage of nighttime prices in some markets, accessing some free electricity at workplaces, and relying on public charging could save PEV drivers even more. It is important to note that PEV drivers experience substantial variation and complexity in energy costs across regions. Even within a given region, there is much local variation because of local rates and special PEV rates offered by the thousands of electric companies in the United States, differences in prices charged at public charging stations, and in some cases free charging at public and work locations.

[7] Much of existing data about PEV driver behavior with respect to electricity prices are shaped by the high income of the initial buyers who are not as sensitive to gasoline or electricity costs as later adopters are likely to be.

Determining Charging Infrastructure Needs

The charging infrastructure is a new part of the vehicle ecosystem that customers must navigate. Potential PEV purchasers need to know what type of charging infrastructure they will need, how to get it installed at home, how to find charging stations when needed, and how to subscribe to or pay for access to the charging stations. Those issues must be considered by potential PEV customers when they consider purchasing a PEV.

Unlike ICE vehicles, for which public fueling stations are the standard, PEVs may be fueled with electricity at home, at workplaces, or at public charging stations (see Chapter 5). In fact, early adopters have primarily satisfied their charging needs at home, and the majority of mainstream PEV adopters are also likely to find home charging to be most convenient. The paradigm for fueling PEVs at the owner's home is a fact not appreciated by many unfamiliar with PEVs, including many policy makers and presumably many potential PEV customers who believe that public charging stations are needed.

Although home charging has been the primary method of refueling, public charging does have an important role to play. The PEV-driver experience is shaped by the presence (availability and visibility) of the charging network in his or her region, and a perception of a lack of public charging infrastructure might hinder PEV deployment. For example, the United States has over 100,000 gasoline stations compared with about 8,400 public charging stations (U.S. Census Bureau 2012). Japan has recognized the importance of public visibility and access to charging and has instituted a major initiative to build an extensive charging infrastructure to instill range confidence and ensure a safety net for limited-range BEV drivers (METI 2010). Drivers of all types of PEVs can use their mobile phones or dashboard displays to navigate and find fueling stations. Apps for PEV owners to monitor their state of charge and to find fueling stations compatible with their vehicles might be particularly important to mitigate consumer concerns about location of fueling stations.

The extent to which customers understand charging infrastructure requirements and needs is unknown; however, it is reasonable to speculate that these considerations are new, and perhaps surprising, to mainstream consumers. The committee notes that the effect of public-charging availability on PEV deployment is not well understood (Lim et al. 2014).

Installing Home Charging

Depending on regional variations, BEV and PHEV buyers might need to choose, acquire, permit, finance, and install a charger for their primary parking location even before purchasing the vehicle. The decision process will require the buyer to understand the differences in charging technologies and possibly to answer the following questions: Do they want or need AC level 1 or level 2 charging? Are upgrades of

BOX 3-1 Calculating Electricity or Fuel Costs for Plug-in Electric and Other Vehicles

People shopping for a vehicle face difficulties in calculating fuel costs per mile, especially if they are trying to compare the fuel costs of vehicles operating on different fuels, including BEVs, PHEVs, HEVs, and ICE vehicles. A typical customer's thought process might proceed as follows:

Possible Cost Calculation for a PEV

Take as an example the five-passenger Nissan Leaf, which gets 3 or 4 miles per kilowatt-hour, depending on speed and the heating or cooling needs of the cabin interior. Assume that the average price of residential electricity in the United States is 12.5 cents per kilowatt-hour; this number is based on a range in the United States of 10 to 15 cents per kilowatt-hour (EIA 2014b), and this in turn translates to 3 to 5 cents per mile. On average, therefore, a Leaf owner who charges at home will pay about 4 cents per mile for electricity (these numbers average local taxes on electricity bills).

Possible Cost Calculation for an ICE Vehicle or an HEV

Gasoline in the United States in August 2014 cost on average about $3.60 per gallon (regional averages ranged from $3.35 in the Gulf Coast region to $3.91 in the West Coast region) (EIA 2014c). An especially efficient HEV, the Prius, gets about 50 miles per gallon. Average ICE passenger vehicles have a fuel economy of 35 mpg. Thus, the Prius would have cost 7.2 cents per mile, and the average passenger vehicle would have cost about 10 cents per mile in the United States in August 2014.

Therefore, in most places in August 2014, BEVs and PHEVs operating in electric mode on stored electricity from the grid cost less than one-half as much per mile as a comparable-sized gasoline vehicle. Specifically, driving 10,000 miles in a gasoline-fueled compact vehicle would have cost around $1,000 for gasoline in 2014; a comparable-sized BEV would have cost less than $500 at that time.

Additional Considerations

- The cost of electricity for a PHEV will vary greatly depending on driving patterns, the charging frequency, and the battery capacity.
- Many PEV drivers might charge away from home, where prices vary. Some PEV drivers might be able to maximize their savings by charging for free at work and getting low off-peak or special PEV rates from their utility.
- Some places, especially California, have tiered rates to discourage high consumption, or time-of-use rates to shift consumption peaks. Those pricing structures can make electricity rates vary for an individual household by time of day, by total monthly consumption, or by climate zone in which the house is located.
- The cost of gasoline can also vary substantially, and that variation complicates the calculation of total fuel costs for PHEV drivers.

household circuits, panels, and even transformers required? How much will the changes cost? What permitting processes, fees, and timing are involved? Will installing a charger require financing (most states require financing of the charger to be separate from that of the vehicle)? How much will the extra cable for 240 V (level 2 charging) cost?[8]

Whether the vehicle is leased or purchased might have an effect on the home-charging decision; people who lease might be less willing to commit to the expense and effort of installing home charging. In other cases, installation concerns might be alleviated if PEV owners can use an existing outlet in their garage. The charging concerns for the 46 percent of new PEV buyers who do not have access to home-charging because they park on the street or live in a multiunit dwelling will be different, but they loom large nonetheless (Axsen and Kurani 2009). Barriers to home-based charging for that market segment are discussed in Chapter 5.

To help mainstream PEV consumers navigate their home-charging needs, some vehicle manufacturers have formed partnerships to streamline the purchase and installation of home chargers. Three examples of partnerships (listed below) are cited in the Federal Highway Administration action plan (FHWA 2012b), but one has been discontinued and another has been reworked.

- *Ford and Best Buy.* Ford initially partnered with Best Buy to offer buyers an integrated process for purchasing a vehicle and installing a home charger; Best Buy's Geek Squad and third-party electrical contractors provided installation services. The charging equipment provided by Leviton could be removed, so that owners could easily take the charger with them when they moved. Ford estimated a cost of around $1,500 for the charging equipment and installation services. The program ended in 2013 when Ford partnered with AeroVironment (Motavalli 2013).
- *General Motors* and *SPX.* General Motors initially offered an AC level 2 home-charging system through a partnership with SPX. The equipment costs were $490; installation costs varied depending on the existing home

[8] The 240 V cables are different from the 110 V cables that come with the vehicle and represent an additional customer expense.

wiring but were typically about $1,500 (GM Authority 2010). General Motors appears to have discontinued its offer for an AC level 2 charging system because the Chevrolet Volt can recharge overnight using an AC level 1 charger.

- *Nissan and AeroVironment*. For the Leaf, Nissan teamed with AeroVironment to provide home charging; Nissan estimates that a private contractor charges about $2,000 on average for a typical installation.

Charging decisions are unique to PEVs and can be overwhelming. Indeed, until the purchase and use process is simpler—for example, a dealer helps the customer manage the whole process—mainstream consumers simply might revert to the more familiar purchase of an ICE vehicle that does not have these added complications (Moore 2014).

Greenness of Plug-in Electric Vehicles

Perceived favorable environmental impact (the greenness) of PEVs motivated some early adopters to purchase PEVs, although environmental impacts appear to be less of a motivator for mainstream market consumers given that just 5 percent of U.S. vehicle purchasers stated a willingness to pay more for an environmentally friendly vehicle (Strategic Vision 2013). Others also find that the impact of a green product on consumer purchases is usually a third trigger, behind price and quality (Esty and Winston 2009). Still, consumers might want to know about the greenness of a PEV—if not for themselves, then when friends, family, and colleagues inquire.

Consumer might ask the following questions: Does driving a PEV actually benefit the environment? Are greenhouse gas emissions and local pollutants decreased if I drive a PEV? Is my electric company a low- or high-carbon emitter? Is my electric company lowering its carbon emissions over time? Similar to computing electricity costs, assessing the greenness of a vehicle is complicated;[9] it includes not only the greenness of the electricity supply used to charge the vehicle but also issues related to how batteries will be disposed of and their contribution to environmental degradation (see Chapter 4 for a discussion of battery recycling). Greenness can be calculated on a well-to-wheels basis, which counts greenhouse gas emissions from a vehicle's tailpipe (tank-to-wheels) and upstream emissions from the energy source used to power a vehicle (well-to-tank).[10] Although the factual details about the cleanness of the electric grid (see Chapter 1) might not be widely known, consumer uncertainty about how green PEVs actually are might cause customers to balk at purchasing one.

[9] Take, for example, the greenness of the electric grid. Depending on whether the power plants in a given area produce electricity from coal, nuclear, wind, hydropower, or other energy source, the greenness can vary greatly.

[10] A more complete analysis of vehicle greenness is a life-cycle assessment that, in addition to the well-to-wheels assessment, takes into account environmental impacts of vehicle production, vehicle use, and disposal of the vehicle at the end of its life.

Lack of Information on Incentives

As discussed in Chapters 2 and 7, the prices of PEVs are higher than those of comparable ICE vehicles. However, various financial incentives for consumers can help offset the difference. PEVs can also have nonfinancial incentives, such as access to high-occupancy-vehicle lanes (see Chapter 7 for an extensive discussion of incentives). Consumer awareness and perceptions of incentives influence their purchase decisions. In Norway and the Netherlands, for example, PEVs are particularly popular because people are aware of and want to take advantage of the generous incentives. In the United States, however, a study by Indiana University shows that 95 percent of the U.S. population in the 21 largest cities is unaware of such incentives (Ingram 2013). A further complication is that federal, state, and municipal incentives are often designed to start and stop at certain times or when certain sales volumes have been achieved. The variability and inconsistency of incentives contribute to customer confusion in evaluating and purchasing PEVs.

One study suggests that the effectiveness of PEV incentives could be enhanced through greater consumer awareness (Krause et al. 2013). Dealers could be a source of information about incentives but are unlikely to have all the necessary information, as discussed below. Moreover, dealers might not want to provide information on incentives for fear of being held accountable if they provide inaccurate information (Cahill et al. 2014). Several Internet sources provide information on incentives, but the degree to which consumers are aware of and use them is unknown.

Lack of Knowledge about the Benefits of Plug-in Electric Vehicles

PEV ownership offers benefits that are familiar to and valued by their drivers but are probably unfamiliar to mainstream consumers. For example, people discover on driving PEVs that they are "peppy" and provide smooth acceleration; moreover, they are quiet (Cahill et al. 2014). In addition, PHEVs do not need oil changes as frequently as ICE vehicles, and BEVs do not require any oil changes (Voelcker 2013b, 2014). Furthermore, regenerative braking and energy recovery, which is novel to many new PEV drivers, provides a unique sensation. Whether engineered as part of the traditional braking system (as in the Toyota Prius) or integrated into the acceleration system (as in the BMW i3 and Tesla Model S), or both, regenerative braking creates a unique driving experience. In contrast to systems that capture kinetic energy when the driver begins to brake, regenerative braking integrated into the acceleration system begins to slow the vehicle and capture energy the moment drivers remove their foot from the gasoline pedal. Some drivers perceive the automatic braking as an advantage, especially in heavy traffic.

Thus, PEVs provide a driving experience that is different from that of a traditional ICE vehicle. Such differences

TABLE 3-5 Consumer Questions Related to Plug-in Electric Vehicle (PEV) Ownership

Vehicle Technology	Vehicle Charging	Other Concerns
Questions PEV Customers Might Have		
Why are PEVs more expensive than conventional vehicles? What is the battery range? How many years will the battery last? Do I need to replace it? Is there a risk of fire? Does the vehicle have sufficient power to drive on the highway? What will be the resale value?	Where can I charge the vehicle? Can I charge it at home? Do I need a special plug? How much will the electricity cost? How long does it take to charge?	What happens if I become stranded because I run out of charge? How green is a particular PEV? Do the batteries end up in landfills?
Questions Customers Might Not Know to Ask		
How much of my battery range will be sacrificed to interior heating and cooling in cold or hot temperatures? Can my regular repair shop perform maintenance and repair work on the vehicle? How does regenerative braking work? Will my battery degrade over time if I use DC fast charging? What are the savings in maintenance and fuel relative to the purchase price of the vehicle?	Where do I charge if I do not have a garage? Are there permitting fees to get a dedicated charger installed in my home? Does my state offer a rebate or incentive to install charging equipment in my home? If I want AC level 2 charging, do I need additional equipment, and how much will it cost? Do I need to inform my utility if I purchase a PEV? Do my rates for charging differ depending on the time of day? If I belong to a charging network, can I use chargers from other networks? How do I find a charging location? Can I reserve a charging location? What if the charger I need to use is being blocked by another vehicle?	How do I file for my tax credit? Does my state offer a tax credit? Do I get free parking? Do I get access to a high-occupancy-vehicle lane with a PEV? Does my employer offer charging at work? Because the car is so quiet, how do I know if it is running?
Questions Friends, Neighbors, and Even Strangers Might Ask		
Did my tax dollars subsidize the purchase of your PEV?	Do these vehicles put excessive demands on the electric grid?	Why do people with PEVs get the most convenient parking spots? Do you pay any fuel fees for highway funds or road taxes?

might appeal more to early adopters than to mainstream consumers, but mainstream consumers will never know whether PEVs meet or even exceed their expectations unless they can drive one, making the test-drive a critically important experience, as discussed later in this chapter.

Summary of Major Perceptual Barriers

From the committee's perspective, the factors discussed above pose major barriers to consumer adoption of PEVs. Confusion continues to loom large in the consumer purchase decision for PEVs. Table 3-5 identifies questions that customers might have when contemplating a PEV purchase. Some questions—Will my battery catch fire? How do I change my battery?—might seem nonsensical to a current PEV owner, but they are questions that consumers have asked and demonstrate the extent of misinformation and the nature of the perceptual barriers that must be overcome before PEV deployment becomes widespread.

Uncertainty and perceived risk plague consumer willingness to purchase innovative products, particularly expensive, long-lived ones, such as vehicles; consumers instead revert to the known and familiar (Mohr et al. 2006). Until they are sufficiently informed and educated, they will likely continue to prefer the relative safety, security, and familiarity of an ICE vehicle. Therefore, mainstream adopters require additional encouragement, information, and incentives to overcome the barriers identified.

Finding: Lack of consumer awareness and knowledge about PEV offerings, incentives, and features is a barrier to the mainstream adoption of PEVs.

Finding: The many perceptual factors that contribute to consumer uncertainty and doubt about the wisdom of a PEV purchase combine with price and range concerns to negatively affect PEV purchases.

VEHICLE DEALERSHIPS: A POTENTIAL SOURCE OF INFORMATION?

A well-known aspect of the vehicle-purchase process entails visits to various dealerships for test-drives and purchase negotiations. Vehicle dealerships traditionally have offered many services to help with the complexity of the purchase process. In fact, in the early years of the automobile market, dealers supported Americans with their vehicle purchases by teaching them to drive and by providing financing, maintenance advice, and service to keep vehicles running. Accordingly, dealers have always played a critical role in the decision-making process of people purchasing a vehicle (Ingram 2014). Given that 56 percent of PEV buyers make over three visits to dealerships, which is twice the number made by non-PEV buyers (Cahill et al. 2014), vehicle dealerships could serve as an important source of information for potential PEV buyers.

Despite the importance of the consumer experience at the dealership, research on dealers and PEVs reveals systemic problems. A Center for Sustainable Energy survey of over 2,000 PEV buyers in California in December 2013 showed that 45 percent of those buyers were "very dissatisfied" and another 38 percent were "dissatisfied" with their purchase experience (Cahill et al. 2014). In the same survey, PEV buyers were asked "how valuable was it to have dealers knowledgeable about various topics." The responses in Table 3-6 show that PEV buyers expect dealer salespeople to be informed about much more than just the vehicle characteristics. However, a *Consumer Reports* mystery shopper recently went to 85 dealerships in four states and found that salespeople were not very knowledgeable (Evarts 2014).

The dealer and salesperson motivation to sell PEVs varies. As noted in the committee's interim report (NRC 2013b), salespeople take three times longer to close a PEV sale than an ICE vehicle sale—time for which they are not differentially compensated. Furthermore, because dealership revenues include charges for after-sales service and support and because PEV maintenance requirements are lower than those for ICE vehicles, that service revenue is missed. Moreover, sales staff at dealerships often turnover rapidly; thus, technically savvy sales staff who are knowledgeable about PEVs are not always available at a given dealer on a given day. Given such turnover, sales training on new products is not always a good investment for the dealership (Darvish 2013).

To address those issues, some dealers in California have hired PEV advocates to sell PEVs specifically. Rather than train the entire sales force, high-volume PEV dealerships have one or two PEV gurus. Moreover, some dealerships now separate floor (personal) sales from Internet sales, and in some situations, 100 percent of PEV sales come from Internet inquiries (UC Davis 2014).[11] The PEV gurus usually are part of the Internet sales team for the dealer; social media are used to steer buyers to those individuals. Partially because dealership salespeople might lack the ability, time, or incentives to educate customers adequately about PEVs, Tesla decided to operate its own dedicated showrooms in which specially trained employees focus exclusively on educating customers about Tesla vehicle ownership. Tesla showrooms are typically styled like boutiques in high-traffic locations, such as a mall, much like Apple stores.

[11] Nissan initially sold Leafs only on the Internet.

In addition to the paucity of knowledgeable salespeople, the *Consumer Reports* study (Evarts 2014) also found that dealers simply did not have PEVs in stock. Only 15 of 85 dealers in four states (California, New York, Maryland, and Oregon)[12] had more than 10 PEVs on their lots; indeed, most dealers had only 1 or 2 PEVs on their lots. That finding is supported by a UC Davis study, which found that 65 percent of California dealerships had no PEVs for sale (UC Davis 2014).

Explanations for the lack of inventory on dealer lots vary. *Consumer Reports* stated that of those with limited or no stock, most (21) dealers attributed limited stock to "high demand;" the next most common explanation, however, was a "lack of consumer interest" in PEVs, also expressed as "nobody buys them." Another possible reason for lack of inventory on dealer lots is based on the financial returns from selling PEVs. If vehicle manufacturers are losing money on PEVs, they could limit availability deliberately (see, for example, Beech 2014; Lutz 2012; Voelcker 2013a; Loveday 2013b). The strategy of limiting inventory further hurts sales and makes it harder to generate economies of scale to drive down manufacturing costs. If production costs are not reduced, prices will remain high.

Additional pressure on dealerships comes from Tesla's challenges to the vehicle dealership franchise laws. In the United States, direct manufacturer-to-consumer vehicle sales are prohibited by franchise laws that require new vehicles to be sold only by licensed, independently owned dealers (Quinland 2013). Such licensed dealers sell new and used cars, including certified preowned vehicles, employ trained vehicle technicians, and offer financing. The National Automobile Dealers Association has challenged Tesla's showroom model in a handful of states (Colorado, Massachusetts, Illinois, New York, Oregon, Texas, and North Carolina).[13] So far, because Tesla does not actually sell vehicles through its showrooms (rather, orders are placed online to the factory), courts have generally upheld its model.[14] Distribution issues are an ongoing area of dispute and although Tesla has advocated for a federal law to overturn the state franchise laws, to date it has been unsuccessful.

Finding: Knowledge of PEVs at dealerships is (at best) uneven and (at worst) insufficient to address consumer questions and concerns.

Finding: Dealers are generally less motivated to sell PEVs than to sell ICE vehicles, and a further complication is that the inventory of PEVs on dealer lots is limited.

[12] Oregon was not named in the article, but personal communication on October 17, 2014, with the author provided the name of the fourth state.

[13] Interestingly, NADA is also starting a communications campaign to counter criticisms that the franchise laws are outdated and that dealers are not willing to change (Nelson 2014).

[14] See, for example, the situation in New Jersey (Friedman 2014; Gilbert 2014).

TABLE 3-6 Ratings of Dealer Knowledge about Various Topics

Topic	Percentage of Respondents Who Assigned the Dealer a Rating of "Very Valuable" (Highest Ranking)
Financial incentives	62
Vehicle performance	62
Nonfinancial incentives	48
Cost of ownership	46
Home charging	42
PEV Smartphone applications	40
Away from home charging	33
Electricity rates for PEVs	32

NOTE: PEV, plug-in electric vehicle.
SOURCE: Based on data from Cahill et al. (2014) and CSE (2014).

STRATEGIES TO OVERCOME BARRIERS TO DEPLOYMENT OF PLUG-IN ELECTRIC VEHICLES

This chapter began by describing diffusion models for new technologies, including PEVs, and then discussed the demographics and behavior of early adopter and mainstream market segments and their implications for adoption and diffusion. A number of barriers consumers face in the PEV purchase process were discussed next. The sections that follow consider ways to address consumer barriers to PEV deployment, including advertising strategies to educate consumers, greater use of Internet resources to disseminate information, and more opportunities for test-drives.

Advertising of Plug-in Electric Vehicles

Given the complexity of the issues consumers face in the PEV purchase decision, the committee examined various information sources that consumers commonly rely on for decision making. One such source is the advertising and marketing of the vehicle manufacturer. Advertising historically has been a way to stimulate interest in new products and to steer customers to dealers. Because advertising and marketing plans are critical aspects of a vehicle manufacturer's strategy, they are proprietary, and the committee did not receive information on individual company efforts, such as how much they spend on advertising to promote PEVs. Although PEVs are advertised in traditional media, casual observation suggests that company efforts to promote PEVs are not nearly as aggressive as their efforts for traditional ICE vehicle makes and models.

Reasons for limited PEV advertising could include the fact that the market is small. A lack of profitability also could be a reason companies do not want to advertise (see, for example, Beech 2014; Lutz 2012; Voelcker 2013a; Loveday 2013b). Companies want to maximize the return from their advertising budgets, and whether the PEV market is sufficiently responsive to warrant larger advertising expenditures is questionable. Regardless of the underlying reasons for what would appear to be a limited effort by any one company to advertise its PEVs, the lack of promotion creates a self-fulfilling prophecy.

One strategy that might be used to overcome barriers to vehicle manufacturers advertising their PEVs is cooperative advertising or joint promotion efforts to communicate the advantages of PEVs. Cooperative advertising is a shared campaign whereby companies work together to achieve an important goal. For example, trade associations for the fishing industry[15] and the recreational-vehicle industry[16] run campaigns to stimulate demand for the product category as a whole. Rather than any one company having to incur the expense of stimulating consumer demand, an industry trade association or other third party undertakes such efforts on behalf of its members. Other examples here include the "Beef: It's What's for Dinner" and "Got Milk" campaigns.

Cooperative advertising campaigns typically exist under at least one of the following conditions:

- The industry as a whole is facing a decline in demand because of competitive threats (when, for example, consumers spend more time with technology gadgets than in going outdoors).
- A new technology is attempting to overcome an incumbent technology, and the combined efforts of the new technology providers might be able to educate consumers in a synergistic fashion.
- The products are commodities; thus, the goods of any one provider are indistinguishable from those of the others (for example, milk).

[15] See, for example, the Recreational Boating and Fishing Foundation (http://takemefishing.org/).

[16] See, for example, Go RVing, Inc. (http://gorving.com/).

Although the third situation does not apply in the automotive industry, the second one is critical. When the motives of an individual stakeholder are insufficient, a collective approach to stimulating awareness and action can be effective. Thus, vehicle manufacturers, suppliers of vehicle components, and charging providers, if united by a strong enough common interest and a capable third-party organization to manage the campaign, could find value in banding together in a manner similar to other industries.

Another partnership that might prove advantageous is between vehicle dealerships and electric utilities. Together they could work to promote the vehicles by emphasizing the convenience and affordability of electric fuel. For example, Austin Energy works with local vehicle dealerships to provide prepaid unlimited public fueling cards for $50 per year (K. Popham, Austin Energy, personal communication, December 18, 2014). The program allows salespeople to offer "free unlimited public charging" on every new vehicle for a year.

In addition to the marketing activities of industry stakeholders, the federal government has played an important role in sponsoring advertising campaigns to support socially beneficial behaviors. The Ad Council (2014), founded in 1941, is a federally subsidized advertising program that partners with national nonprofits or federal agencies on multimedia marketing campaigns. It selects important public issues (such as the Smokey the Bear "Only You Can Prevent Forest Fires," the United Negro College Fund "A Mind Is a Terrible Thing to Waste," and "Friends Don't Let Friends Drive Drunk"), partners with a sponsoring organization, and then stimulates action on those issues through advertising programs. Partner organizations (such as the Department of Energy [DOE] in the case of PEVs) are considered the issue experts and, as such, sponsor the campaign and are responsible for production and distribution costs (research, multimedia production, multimedia distribution, social media, and public relations); the media space or air time is donated. The Ad Council asks for a campaign commitment of at least 3 years, which is consistent with models of consumer learning and engagement for risky, durable purchases. For an Ad Council campaign, DOE could work with marketing experts to craft appropriate messaging, including accurate information about federal tax credits and other incentives; the value proposition for PEVs generally, including lower operating costs; and the identification of people who could usefully own PEVs. More broadly, the government's objectives (energy security and clean transportation) could also be part of the message.

Finding: The federal government has a mechanism to communicate messages to the general public for issues deemed to be in the public interest.

Recommendation: To provide accurate consumer information and awareness, the federal government should make use of its Ad Council program, particularly in key geographic markets, to provide accurate information about federal tax credits and other incentives, the value proposition for PEV ownership, and who could usefully own a PEV.

Internet Resources for Information on Plug-in Electric Vehicles

For the motivated and savvy consumer, a plethora of online resources are available to research PEVs (see Table 3-7) and the other components of the purchase decision. Online research can provide make and model availability, prices, technical specifications and reviews; describe the charging infrastructure, including locations of public charging stations; list incentives by state or zip code; and even give estimates of total cost of ownership. Traditional car-buying websites, such as Kelley Blue Book and Edmunds, have areas dedicated to PEVs. Manufacturers of PEVs have information on their websites. Many automotive enthusiasts also provide information on various other websites. Because of the importance of electricity and charging to their business models, most utilities have a section of their websites dedicated to PEVs. Many nonprofit environmental organizations have sections for PEVs. Consumers looking for information on charging infrastructure specifically can find many resources through the various private companies offering public charging. Finally, federal and state websites offer useful resources, including calculators for electricity costs.

The plethora of online information provides an opportunity to overcome the lack of consumer awareness and knowledge about PEVs, but two potential problems arise. First, the sheer number of Web resources might cause consumers to become overwhelmed and confused. Studies of consumer decision making show that information overload is negatively associated with purchase (Herbig and Kramer 1994); too many options create confusion (Schwartz 2005; Scheibehenne et al. 2010). Despite the wealth of information or perhaps because of it, consumer knowledge about PEVs is not as great or as sophisticated as it could be, and misperceptions certainly continue to exist.

Second, finding an easy-to-use source of credible, reputable information can be difficult. For example, an online search to find information related to purchasing a PEV yields a wide array of links, such as sponsored advertisements for PEVs, vehicle-manufacturer websites, news articles about PEVs, blog posts from PEV enthusiasts, buyer guides, information from nonprofits encouraging PEVs, information on tax credits, and even paid Google AdWords campaigns for fuel-efficient ICE vehicles and technologies.[17] The confusing array of results—including misinformation on PEVs—emphasizes the need for a central, credible (unbiased), easy-to-use resource to simplify consumer information needs.

If consumers are lucky, they will find the useful federal government websites for PEVs. The Alternative Fuels

[17] Search results on any given day and computer are conditioned by the cookies on an individual user's computer, search engine marketing at the time, and other factors.

TABLE 3-7 Websites with Information on Plug-in Electric Vehicles

Category	URL	Type of Information Available
Vehicle reviews	http://www.edmunds.com/hybrid/ http://www.kbb.com/electric-car/?vehicleclass=newcar&intent=buy-new&filter=hasincentives http://www.consumerreports.org/cro/cars/hybrids-evs.htm http://www.cars.com/guides/all/all/?prop63=Electric%20Powered&highMpgId=1836&sf1Dir=ASC	Reviews, technical specifications, make and model availability
Vehicle industry blogs and websites	http://www.greencarreports.com/ http://www.epri.com/Our-Work/Pages/Electric-Transportation.aspx http://www.electrificationcoalition.org/ http://www.plugincars.com/ http://www.howtoelectriccar.com/is-an-electric-car-right-for-me/ https://www.aepohio.com/save/ElectricVehicles/EVRight.aspx http://www.electricdrive.org/ http://www.electriccarbuyer.com/guide/ http://insideevs.com/ http://www.pluginamerica.org/ http://driveelectricweek.org/ http://green.autoblog.com/ http://evsolutions.avinc.com/electric_vehicles/ http://cleantechnica.com/category/clean-transport-2/electric-vehicles/ http://chargedevs.com/ http://www.thecarconnection.com/category/new,electric-car http://www.huffingtonpost.com/news/electric-cars/ http://www.tva.com/environment/technology/electric_transportation.htm https://www.alamedamp.com/types-of-electric-vehicles http://transportevolved.com/	Market trends, including sales volumes, PEV news, reviews
Nonprofit organizations	http://www.nrdc.org/energy/vehicles/green-car-tech.asp http://www.edf.org/transportation/fuel-economy-standards http://content.sierraclub.org/evguide/	Environmental impacts of PEVs, incentives, policy, dispelling myths
Charging-infrastructure locators	http://www.plugshare.com/ http://www.afdc.energy.gov/fuels/electricity_infrastructure.html http://www.nrgevgo.com/ http://www.chargepoint.com/ www.juicebarev.com	Maps and search tools to find charging infrastructure, availability of chargers, subscription plans
Cost of ownership calculators	http://www.afdc.energy.gov/calc/ http://energy.gov/maps/egallon http://www.electrificationcoalition.org/sites/default/files/EC_State_of_PEV_Market_Final_1.pdf	Calculators for cost of ownership of PEVs based on local and individual variables
Federal government resources	http://avt.inel.gov/ http://avt.inel.gov/hev.shtml www.fueleconomy.gov http://energy.gov/maps/egallon http://www.evroadmap.us/ http://www.afdc.energy.gov/vehicles/electric.html http://www1.eere.energy.gov/cleancities/ http://energy.gov/eere/vehicles/vehicle-technologies-office-hybrid-and-vehicle-systems http://energy.gov/eere/vehicles/vehicle-technologies-office-information-resources http://energy.gov/eere/vehicles/vehicle-technologies-office-ev-everywhere-grand-challenge http://www.epa.gov/greenvehicles	Incentive information, regulation information, data on PEVs, government research, and deployment initiatives
State government resources	https://energycenter.org/ http://www.westcoastgreenhighway.com/electrichighway.htm http://www.in.gov/oed/2675.htm http://www.plugandgonow.com/	State-specific incentives and policies, consumer guides, resources for advocates, state, local and regional charger maps

Data Center (DOE 2013a) provides comprehensive information on vehicle and fuel characteristics and infrastructure and useful fuel cost calculators (DOE 2013b, 2014a). The DOE Energy Efficiency and Renewable Energy Clean Cities website (DOE 2014b) also provides valuable information specific to localities and regions where a Clean City Coalition operates. Unfortunately, those websites do not appear consistently high in search results for PEV information. Moreover, compared with other shorter and catchier Web addresses—such as greencarreports.com, plugincars.com, and plugandgonow.com—consumers might find it difficult to remember the URLs for the government sites (see Table 3-7). The committee did not have access to data on the extent to which car shoppers relied on government website resources.

Furthermore, given the lack of evidence on how consumers use objective information (versus their perceptions) in purchase decisions, the potential effect of calculators for fuel cost and total cost of ownership on a customer's evaluation of PEVs is unknown. One common strategy used to evaluate the responsiveness of website visitors to various types of and formats for online information is called A-B testing. A-B testing presents version A of the information for a period of time and tracks visitor activity and then presents version B for a similar period of time. The two information strategies are then compared to reveal the differential impact of the information presentation.

Finding: Government websites provide useful information for motivated PEV shoppers; however, the degree to which they are easy to find, remember, and share is unknown, as is their actual impact on consumer perceptions and behavior.

Recommendation: The federal government should engage a knowledgeable, customer-oriented digital marketing consultant to market its online resources and then evaluate their impact. Marketing activities could include purchasing a user-friendly, memorable domain name, running various A-B tests, optimizing search engine marketing to allow shoppers to find useful resources more easily, using sharing tools to facilitate dissemination among online networks, and identifying key partners to use application protocol interfaces to promote greater consistency of information.

Test-Drive Events and Regional Experience Centers

Test-drives are critically important for potential PEV buyers because they allow customers to assess the driving characteristics of PEVs.[18] Because driver experience with PEVs is a critical aspect of the purchase decision, vehicle manufacturers, vehicle dealers, nonprofit organizations, and various DOE initiatives have experimented with a variety of events to draw customers to experience PEVs. For the EV1 launch, GM took it to several U.S. cities for month-long tests. In 2012, it offered 3-day test-drives of its Chevrolet Volt in major cities. More recently, Fiat took its 500e BEV to 30 corporate campuses in California and offered lunch and test-drives to employees (Anders 2012). Plug-in America has been organizing National Drive Electric Week (formerly National Plug-in Day) since 2011. In 2013, 80 events sponsored by corporations, nonprofits, and PEV enthusiasts across the country hosted over 33,000 participants and gave over 2,700 test-drives (Plug-in America 2013).

DOE recognized the importance of consumer demonstrations in its July 2014 call for proposals through the Clean Cities program (DOE 2014c). It is offering funding for 7 to 15 deployment projects in three areas: on-the-road demonstrations, safety-related training, and emergency preparedness. On-road demonstrations will allow people to have first-hand PEV experience for extended periods of time. Whether the experience is through car sharing, rental car, or commercial fleet leasing programs, more drivers will understand the benefits of PEVs and be more prepared to evaluate them knowledgeably and perhaps more likely to purchase them.

Vehicle manufacturers—including Cadillac, BMW, and Porsche—also are developing regional experience centers (Colias 2014). To adapt to shifting shopping habits, vehicle manufacturers are offering customers an opportunity to look at vehicles in a less sales-oriented environment. For example, "BMW's new retail sales model includes plans for regional pools of test cars with a wider range of models, giving dealers access to more demo models than any store could stock" (Colias 2014). Because the regional facilities will supplement, not supplant, the existing dealer networks and because they address a different point in the consumer decision-making journey before the actual purchase decision, the regional centers are (so far, at least) not in conflict with dealer franchise laws.

The committee finds that such regional centers could be a useful strategy to help mainstream customers gain more hands-on experience with PEVs. For customers who want to compare and contrast different types of PEVs, doing so at a central location would be much easier than having to visit three or four dealerships, especially given that dealer salespeople might not be as knowledgeable as desired and given the dearth of PEV models available on the lots. There might also be a business model whereby vehicle manufacturers hire a third party to provide ride-and-drive opportunities at workplaces and community events.

Finding: The test-drive experience, including an opportunity to become familiar with vehicle range and charging, is a critical aspect of the consumer decision-making process for PEVs. Thus, more initiatives that offer "ride and drives" for a range of PEVs at a single location would be helpful.

[18] The committee notes that a test-drive will not allow the driver to experience the effect of driving factors on range and most likely will not provide an opportunity to recharge the vehicle. One could argue that drivers also do not experience the true range or refueling of an ICE vehicle during a test drive. However, ICE-vehicle buyers have enough experience to make an informed decision about those topics to alleviate concern. Potential PEV buyers, on the other hand, will likely lack information on those topics and will have to trust the information provided by the dealer.

Recommendation: The federal government should explore opportunities for a vehicle-industry effort to provide a regional PEV experience center to provide important test-drive opportunities.

Other Opportunities to Experience Plug-in Electric Vehicles

Another opportunity for people to experience PEVs can come from fleets. For example, some municipalities in Japan allowed citizens to reserve a city-owned PEV to drive on weekends (H. Matsuura, Kanagawa Prefectural Government, personal communication, December 11, 2013). Employers that have PEVs in their corporate fleets could consider a similar idea, either as a perk for employees or as a way to promote an environmental image. Rental car fleets could also provide an opportunity for customers to experience PEVs. Hertz has its "green collection," which allows drivers to experience PHEVs (not BEVs).

As noted previously, car sharing is a growing trend in the vehicle market, particularly in large cities, where personal vehicle ownership is less necessary and less convenient. PEVs seem like a good choice of vehicle for many car-sharing enterprises given the often short distances traveled per rental and the environmental values that motivate some car sharers. However, companies that want to use PEVs in car-sharing fleets face barriers, such as vehicles that might be more expensive and have a limited range, which might make them inconvenient for customers. The companies will also need charging stations and creative strategies for managing the operation of the fleets.[19]

A successful car-sharing program that uses PEVs has been implemented in Madrid, where Respiro Car Sharing, Nissan Leaf, and NH Hotels have collaborated to develop a sustainable mobility plan for the city, where many vehicles travel fewer than 50 km per day (EnergyNews 2013). In Paris, AutoLib was introduced in 2012 with 250 vehicles and 250 stations. Eventually, the company plans to grow to 3,000 vehicles and 1,000 stations at an investment cost of €235 million. PEV car sharing has also been successful in urban centers in places like in the Netherlands, where there is a scarcity of parking spaces and having a reserved PEV parking location is valuable. Car2Go has three all-BEV fleets worldwide (San Diego, Amsterdam, and Stuttgart); its 27 other locations offer PEVs and gasoline vehicles. And BMW has its i-Drive initiative in the San Francisco Bay Area, where it has a point-to-point service similar to Car2Go. Members of the service can access a car from one of the multiple locations in the cities where the service is available and can then leave the car at the destination where they want to go. Other members find the car through radio-frequency identification technology and use it if the location fits their needs.

To the extent that car-sharing fleets use PEVs, they also represent a way for the public to experience PEVs and might represent an important means for introducing PEVs to market segments that might not be traditional new-vehicle buyers. For example, Car2Go indicated that its customers tend to be young, low-to-moderate income earners, often students, and urban dwellers who live in areas that have high congestion and limited parking (Cully 2014).

FEDERAL GOVERNMENT EFFORTS TO FAMILIARIZE CONSUMERS WITH PLUG-IN ELECTRIC VEHICLES: CLEAN CITIES COALITION

DOE's major effort in PEV deployment is the Clean Cities program, which is managed by the the department's Vehicle Technology Office and stems from the Energy Policy Act of 1992. The program goal is to support local actions for reducing petroleum use in transportation by promoting alternative-fuel vehicles. Over 80 local Clean Cities coalitions represent about 80 percent of the U.S. population (Frades 2014). In 2011, an additional funding stream of $8.5 million was allocated for 16 PEV community-readiness projects to support public-private partnerships in deployment of PEVs and their associated infrastructure. Figure 3-6 shows the Clean Cities coalitions that received funding for PEV readiness.

Although the coalitions act locally, one of the most useful and comprehensive resources for PEV owners and policy makers from the Clean Cities program is the DOE Vehicle Technology Office Alternative Fuels Data Center website in Table 3-7. Particularly useful is the station locator (DOE 2014a), which allows searching for PEV charging by location, by charging technology, by station type, and by payment method accepted.

Although PEVs are only a small part of the Clean Cities initiative, the 2011 Clean Cities strategic plan describes the key areas for PEV deployment efforts, including strong coalitions and partnerships, infrastructure deployment in vehicle-manufacturer target markets, information provision and data collection, and training (DOE 2011). In their readiness plans, the 16 Clean Cities coalitions identified barriers to PEV adoption and infrastructure deployment and implemented plans to overcome the barriers. One of the primary barriers identified was lack of awareness of and information about PEVs on the part of many stakeholders, including vehicle purchasers and the government. To overcome the barriers, the coalitions supported outreach, education, training, and marketing efforts, including hosting events for people to experience PEVs. They also produced templates, guides, and tools for outreach to the public, businesses, and local government officials (Frades 2014). The committee found that Clean Cities coalitions were vital to increased deployment

[19] As noted, car-sharing programs are allowing point-to-point rides, but that freedom also makes managing an all-BEV fleet difficult because members could drain the battery and leave a vehicle stranded. Given the few public charging stations in San Diego, Car2Go has managed an all-BEV fleet by building and operating a charging barn where it charges its entire fleet on average every 2 days.

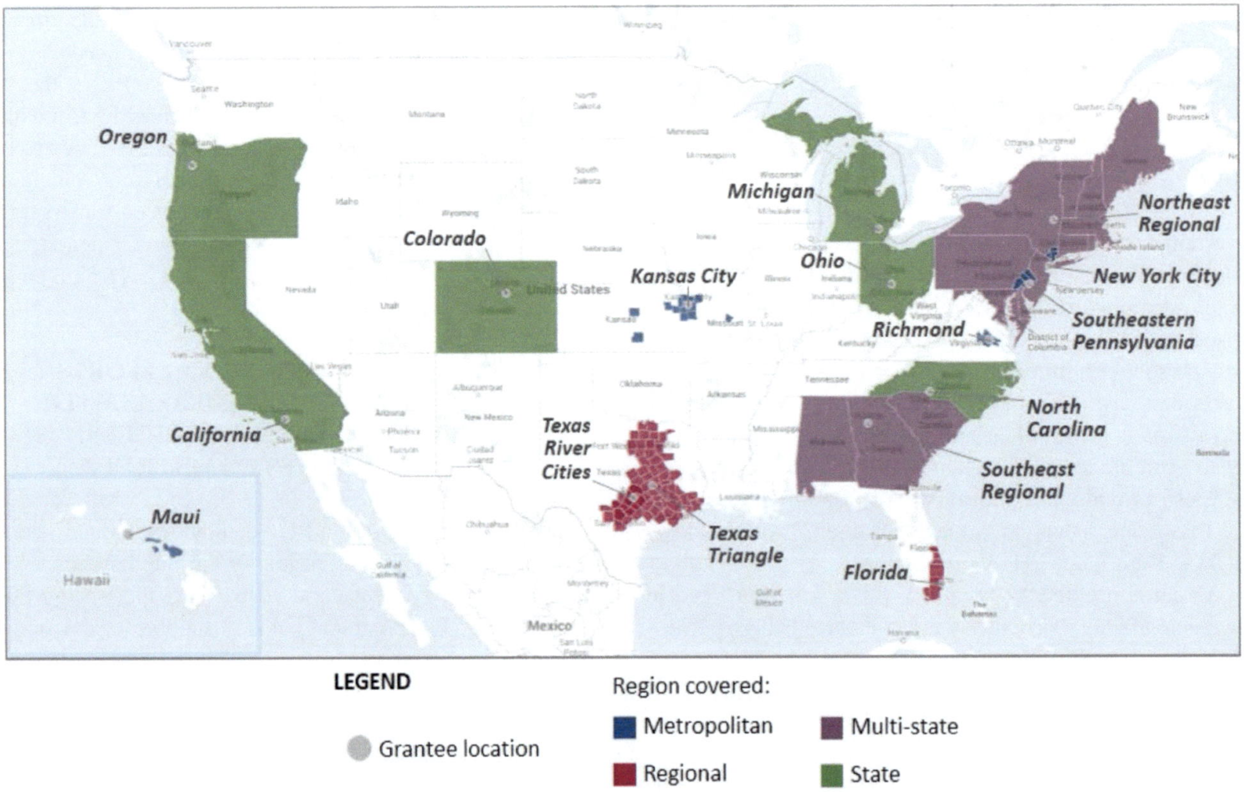

FIGURE 3-6 Clean Cities coalitions funded for community-readiness and planning for PEVs and PEV charging infrastructure. The grants are to Clean Cities coalitions with a focus on PEV deployment. SOURCE: Frades (2014).

in some places, such as Atlanta. Aside from DOE's research funding, the Clean Cities coalitions that are working on PEV deployment represent DOE's most prominent efforts in PEV deployment.

Finding: Clean Cities coalitions have been vital to increased deployment in some localities and represent DOE's most prominent efforts in PEV deployment aside from its research funding.

FLEET PURCHASES

One method to increase PEV deployment is purchasing them for fleets, particularly fleets where vehicles leave and return to the same base and have similar daily routes. Vehicle fleet sales make up 20-22 percent of the U.S. market (Automotive Fleet 2013). The exact size of the fleet market is hard to measure because not all purchasers identify themselves as a business purchaser, and some fleet vehicles are driven for private use. Figure 3-7 shows that the fleet category includes an array of buyers, including rental companies, which account for over 80 percent of fleet purchases. Governments comprise the smallest category of fleet buyers, about 4.1 percent. Fleet managers are looking to alternative-fuel vehicles, including PEVs, to meet societal responsibilities to lower greenhouse gas emissions, to lower fuel and operating costs, and to maintain an environmentally friendly image.

As Table 3-8 shows, information resources for fleet managers who are tasked with greening their fleets are plentiful. However, PEV deployment in fleets has not been strong. Accordingly, this final section provides a brief overview of the three main classes of fleet buyers and an assessment of the barriers to and opportunities for facilitating PEV deployment in this segment.

Rental Fleets

Although rental companies comprise the largest fleet buyers, the viability of PEVs in their fleets is constrained by not knowing a typical customer's driving range and the need for charging and the difficulty of gauging the resale value of a PEV (El-Moursi 2013). Rates for renting PEVs are generally higher than for conventional vehicles, and their availability is harder to ascertain. When coupled with uncertainties about charging and how far customers will drive the rental vehicle, the business proposition for PEVs in rental fleets is unclear.

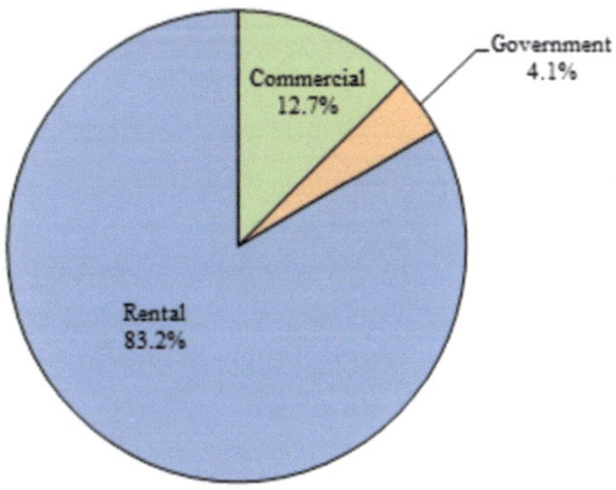

FIGURE 3-7 Fleet sales for passenger vehicles for 2012 by fleet purchase agency. SOURCE: Based on data from Automotive Fleet (2013).

TABLE 3-8 Information Resources for Fleet Managers

Resource	URL
EV Solutions: Operation Audits	http://evsolutions.avinc.com/yourbusiness/fleets/gov
Vehicle Trends & Maintenance Costs Survey	http://www.fleetanswers.com/sites/default/files/Dow%20Kokam%20 Survey%20Report_0.pdf
Plug-In Vehicle Strategic Planning/Feasibility Study Template, Denver Metro Clean Cities Coalition	http://www.denvercleancities.org/Plug-In%20Vehicle%20Assessment-%20Denver%20Metro%20Clean%20Cities%20with%20logo.pdf
Demand Assessment of First-Mover Hybrid and Electric Truck Fleets, CALSTART	http://www.calstart.org/Libraries/Publications/Demand_Assessment_of_First-Mover_Hybrid_and_Electric_Truck_Fleets.sflb.ashx
Fleet Electrification Roadmap, Electrification Coalition	http://www.electrificationcoalition.org/sites/default/files/EC-Fleet-Roadmap-screen.pdf
PG&E in California: PEV Case Study: It's Electrifying: Positive Returns in PEV Deployment, Electrification Coalition	http://fleetanswers.com/sites/default/files/PGE%20case%20study%20Final.pdf
FedEx: EV Case Study, The Electric Drive Bellwether?, Electrification Coalition	http://www.fleetanswers.com/sites/default/files/FedEx_case_study.pdf
Joint Procurement of EVs and PHEVs in Sweden, Clean Fleets	http://www.clean-fleets.eu/fileadmin/files/CF_case_study_sweden_04.pdf

Despite the uncertainty, Hertz is experimenting with PEVs in its rental fleets in key locations and where partners, such as nearby hotels, are equipped to address charging (Hidary 2012).

Corporate or Business Fleets

General Electric made big news when it announced in 2010 that it would convert half of its fleet to PEVs (25,000 vehicles), of which one-half would be Chevrolet Volts (Richard 2010; Antich 2011). Many other companies, including Pepsi, Frito-Lay, and Cisco, have also stated objectives to green their fleets. For business fleets, issues related to limited choice of models, charging infrastructure, and higher initial prices compared with ICE vehicles pose barriers to adoption by fleets. Furthermore, fleet managers face challenges in trying to manage routes (Westervelt 2012; Hanson 2013). Unlike consumers, who do not appear to consider total cost of ownership when deciding whether to purchase a PEV, fleet managers attend carefully to such issues. As noted by Wolski (2013), "the real tipping point [for broad implementation of PEVs in commercial fleets] is when the total operating costs plus the capital costs balance out in three years or less," an unlikely scenario for PEVs in the near term.

Electric utilities across the country provide an interesting opportunity for PEV fleet deployment. Given that many utilities are actively working with the vehicle industry in PEV deployment, they should be one of the main fleet owners transitioning to PEVs. Their lessons could help to inform other fleet managers. The DOE Clean Cities program discussed above includes incentives to convert business fleets and offers information for fleet managers (DOE 2012); these resources will need to be updated as PEV deployment occurs and lessons are learned.

Government Fleets

The federal government has a vehicle fleet comprised of more than 600,000 vehicles and is, therefore, the nation's single largest fleet operator (GSA 2011). The General Services Administration procures about 65,000 vehicles each year and owns and leases about 210,000 vehicles to federal agencies. State, county, and municipal governments also have their own fleets.

In 2011, President Obama ordered that by the end of 2015, all new light-duty vehicles purchased or leased by federal agencies be alternative-fuel vehicles, which include flex-fuel vehicles, HEVs, PEVs, compressed natural gas and biofuel vehicles (Obama 2011). The order also encouraged the agencies to support the development of alternative-fueling infrastructure. Through 2013, the majority of alternative-fuel vehicles purchased by federal fleets have been flex-fuel vehicles that can operate on E-85 rather than vehicles fueled by electricity, natural gas, or other fuels (GSA 2013). Specifically, about half of reported federal fleet purchases in 2013 were flex-fuel vehicles, and 36 percent were conventional gasoline vehicles, some of which might even satisfy the mandate in areas where alternative fuels are not considered to be available. Few PHEVs or BEVs have been purchased; PEVs represent about 1.2 percent of reported federal fleet vehicle purchases. DOE itself has purchased few PEVs; only 0.73 percent of its fleet are PEVs.

Like individual consumers, the government faces barriers in adopting PEVs. The price of PEVs is a particularly high barrier for the federal agencies; tax incentives available to consumers are not available for government fleets. Another barrier is that vehicle purchases come from the capital budgets, and fuel expenses come from operating budgets (DOE 2014d). The need to provide charging infrastructure at its fleet facilities poses yet another barrier. Finally, government procurement practices have been described as excessively complicated and lead some to wonder whether government fleet sales are a realistic way to demonstrate the suitability of PEVs in fleets.

The committee notes that although the total number of vehicles in government fleets is small compared with the total number of vehicles in the overall market, converting some portion of the fleets to PEVs is important. First, people expect leadership from their government. Given the mandates for energy security and clean transportation—the very motivations for this committee's work—the symbolic importance of the government's own efforts lend authenticity to the mandates. Second, the large number of people working at all levels of government, particularly in the federal government, could play a role in information diffusion and the education of friends and neighbors. Third, given that DOE is the main government agency working to deploy PEVs, it should serve as a model by deploying PEVs in its own fleets. To explore ways to remove barriers to PEV deployment across the private sector while not removing barriers in its own organization is poor policy.

Recommendation: To lend authenticity to the federal government's initiative and to enhance the visibility of PEVs generally, the federal government should demonstrate leadership by adding PEVs to its fleets and offering charging infrastructure at its facilities.

Recommendation: DOE should itself serve as a model by adding PEVs to its fleets and use its experience to discern best practices for dissemination to the private sector and other government fleets.

REFERENCES

Ad Council. 2014. "Become an Ad Council Campaign." http://www.adcouncil.org/Working-With-Us/Become-an-Ad-Council-Campaign. Accessed October 23, 2014.

Adner, R. 2006. "Match Your Innovation Strategy to Your Innovation Ecosystem." Harvard Business Review, R0604F-PDF-ENG, April. http://hbr.org/2006/04/match-your-innovation-strategy-to-your-innovation-ecosystem/ar/1.

Aggarwal, P., T. Cha, and D. Wilemon. 1998. Barriers to the adoption of really-new products and the role of surrogate buyers. Journal of Consumer Marketing 15(4): 358-371.

Anders, G. 2012. "Test Drive a Volt, Please!—How GM's Pitch Backfired." Forbes, July 18. http://www.forbes.com/sites/georgeanders/2012/07/18/volt-test-drive/.

Antich, M. 2011. "Half of GE's 30,000-Unit Fleet to be Electric Vehicles by 2015." Automotive Fleet, February. http://www.automotive-fleet.com/channel/leasing/article/story/2011/02/half-of-ge-s-30-000-unit-fleet-to-be-electric-vehicles-by-2015.aspx.

Automotive Fleet. 2013. "J.D. Power and LMC Automotive say U.S. Fleet Sales Improve in September."September 19. http://www.automotive-fleet.com/channel/vehicle-research/news/story/2013/09/j-d-power-and-lmc-automotive-say-u-s-fleet-sales-improve-in-september.aspx.

Axsen, J., and K.S. Kurani. 2009. Early U.S. market for plug-in hybrid electric vehicles: Anticipating consumer recharge potential and design priorities. Transportation Research Record 2139: 64-72.

Beech, E. 2014. "Chrysler CEO: Please Don't Buy Fiat 500e Electric Car." Reuters, May 21. http://www.reuters.com/article/2014/05/21/chrsyelr-ceo-evs-idUSL1N0O71MS20140521.

Brook, D. 2014. "Should You Become a Carsharing Operator?" Metavera, March 13. http://www.slideshare.net/Metavera/should-you-become-a-car-sharing-operator-introduction.

Brown, S., D. Pyke, and P. Steenhof. 2010. Electric vehicles: The role and importance of standards in an emerging market. Energy Policy 38(7): 3797-3806.

Cahill, E., J. Davies-Shawhyde, and T. Turrentine. 2014. "Zero-emission Vehicles and Retail Innovation in the U.S. Automotive Sector: An Exploration of the Consumer Purchase Experience for Plug-in Electric Vehicles." University of California, Davis Institute of Transportation Studies Working Paper, August 2014.

Caperello, N., J. Tyree-Hageman, and K. Kurani. 2014. "Engendering the Future of Electric Vehicles: Conversations with Men and Women." University of California, Davis Institute of Transportation Studies Working Paper UCD-ITS-WP-14. http://www.its.ucdavis.edu/research/publications/publication-detail/?pub_id=2095.

CNCDA (California New Car Dealers Association). 2015. California Auto Outlook: Fourth Quarter 2014. Volume 11, Number 1, February. http://www.cncda.org/CMS/Pubs/Cal_Covering_4Q_14.pdf.

Cobb, J. 2013. "December 2012 Dashboard." Hybrid Cars, January 8. http://www.hybridcars.com/december-2012-dashboard.

Cobb, J. 2015. "December 2014 Dashboard." Hybrid Cars, January 6. http://www.hybridcars.com/december-2014-dashboard/.

Colias, M. 2014. "Cadillac's European Retail Strategy May Be Adapted for U.S." Automotive News, March 24. http://www.autonews.com/article/20140324/RETAIL/303249978/cadillacs-european-retailstrategy-may-be-adapted-for-u.s.

Consumers Union and the Union of Concerned Scientists. 2013. Electric Vehicle Survey Methodology and Assumptions: American Driving Habits, Vehicle Needs, and Attitudes Towards Electric Vehicles. December. http://www.ucsusa.org/assets/documents/clean_vehicles/UCS-and-CU-Electric-Vehicle-Survey-Methodology.pdf.

CSE (Center for Sustainable Energy). 2014. "California Air Resources Board Clean Vehicle Rebate Project, EV Consumer Survey Dashboard." http://energycenter.org/clean-vehicle-rebate-project/survey-dashboard.

Cully, M. 2014. Presentation to the Committee on Overcoming Barriers to Electric-Vehicle Deployment, Washington, DC, May 6.

Darvish, T. 2013. Presentation to the Committee on Overcoming Barriers to Electric-Vehicle Deployment, Washington, DC, December 3.

Dediu, H. 2012. "When Will Smartphones Reach Saturation in the US?" Asymco, April 11. http://www.asymco.com/2012/04/11/.

DOE (U.S. Department of Energy). 2011. Clean Cities Five-Year Strategic Plan: Reducing Petroleum Dependence in On-Road Transportation in the United States. https://www1.eere.energy.gov/cleancities/pdfs/strategic_plan.pdf.

DOE. 2012. Plug-in Electric Vehicle Handbook for Fleets. Alternative Fuels Data Center. http://www.afdc.energy.gov/pdfs/pev_handbook.pdf.

DOE. 2013a. "Alternative Fuels and Advanced Vehicles." Alternative Fuels Data Center. http://www.afdc.energy.gov/fuels/.

DOE. 2013b. "Vehicle Cost Calculator." Alternative Fuels Data Center. http://www.afdc.energy.gov/calc/.

DOE. 2014a. "Alternative Fueling Station Locator." Alternative Fuels Data Center. http://www.afdc.energy.gov/locator/stations/.

DOE. 2014b. "Building Partnerships to Reduce Petroleum Use in Transportation." http://www1.eere.energy.gov/cleancities/.

DOE. 2014c. "$4.5 M in Funding Available for Alternative Fuel Market Readiness." http://www1.eere.energy.gov/cleancities/news_detail.html?news_id=21555.

DOE. 2014d. Presentation at the NEXTSteps Symposia, University of California at Davis.

Edmunds. 2013. 4th Quarter-2013 Used Market Quarterly Report. http://www.edmunds.com/industry-center/data/used-car-market-quarterly-report.html.

Egbue, O., and S. Long. 2012. Barriers to widespread adoption of electric vehicles: An analysis of consumer attitudes and perceptions. Energy Policy 48: 717-729.

EIA (U.S. Energy Information Administration). 2014a. Annual Energy Outlook 2014 with Projections to 2040. Washington, DC: U.S. Department of Energy, April. http://www.eia.gov/forecasts/aeo/pdf/0383(2014).pdf.

EIA. 2014b. "Table 5.6.A. Average Retail Price of Electricity to Ultimate Customers by End-Use Sector." Electric Power Monthly. http://www.eia.gov/electricity/monthly/epm_table_grapher.cfm?t=epmt_5_6_a. Accessed December 30, 2014.

EIA. 2014c. "Petroleum and Other Liquids: Weekly Retail Gasoline and Diesel Prices." http://www.eia.gov/dnav/pet/pet_pri_gnd_a_epm0_pte_dpgal_m.htm.

El-Moursi, S. 2013. "Rent Green: Will Electric and Hybrid Car Rentals Take Off?" Buzz: The Power2Switch Blog, March 1. https://power2switch.com/blog/rent-green-will-electric-and-hybrid-car-rentals-take-off/.

EnergyNews. 2013. "Madrid City Council Goes for Car Sharing." http://www.energynews.es/english/nh-hoteles-se-apunta-a-la-movilidad-el-ctrica-1365489521/.

EPA (U.S. Environmental Protection Agency). 2014. Trends Report. Light-Duty Automotive Technology, Carbon Dioxide Emissions, and Fuel Economy Trends: 1975 Through 2014. EPA-420-R-14-023. October. http://epa.gov/fueleconomy/fetrends/1975-2014/420r14023.pdf.

EPA/NHTSA (U.S. Environmental Protection Agency and National Highway Traffic Safety Administration). 2012. "2017 and Later Model Year Light-Duty Vehicle Greenhouse Gas Emissions and Corporate Average Fuel Economy Standards." Federal Register 77(199): October. http://www.gpo.gov/fdsys/pkg/FR-2012-10-15/pdf/2012-21972.pdf.

Eppstein, M.J., D.K. Grover, J.S. Marshall, D.M. Rizzo. 2011. An agent-based model to study market penetration of plug-in hybrid electric vehicles. Energy Policy 39(6): 3789-3802.

Esty, D., and A. Winston. 2009. Green to Gold: How Smart Companies Use Environmental Strategy to Innovate, Create Value, and Build Competitive Advantage. Hoboken, NJ: Wiley.

Evarts, E. 2014. Dealers not always plugged in about electric cars, Consumer Reports' study reveals. Consumer Reports, April. http://www.consumerreports.org/cro/news/2014/04/dealers-not-always-plugged-in-about-electric-cars-secret-shopper-study-reveals/index.htm.

FHWA (Federal Highway Administration). 2011. Summary of Travel Trends: 2009 National Household Travel Survey. FHWA-PL-11-022. U.S. Department of Transportation. http://nhts.ornl.gov/2009/pub/stt.pdf.

FHWA. 2012a. "Public Road Mileage-VMT-Lane Miles 1920-2012." Office of Highway Policy Information Highway Statistics. http://www.fhwa.dot.gov/policyinformation/statistics/2012/vmt421c.cfm.

FHWA. 2012b. "An Action Plan to Integrate Plug-in Electric Vehicles with the U.S. Electrical Grid." http://www.fhwa.dot.gov/environment/climate_change/energy/publications_and_tools/pev_action_plan/index.cfm.

FHWA. 2014. "New Data Show U.S. Driving at Highest Level in Six Years." https://www.fhwa.dot.gov/pressroom/fhwa1430.cfm.

FHWA. 2015. "Highway Statistics 2013. Annual Vehicle Distance Traveled in Miles and Related Data-2013-by Highway Category and Vehicle Type." Office of Highway Policy Information. http://www.fhwa.dot.gov/policyinformation/statistics/2013/vm1.cfm.

Frades, M. 2014. A Guide to the Lessons Learned from the Clean Cities Community Electric Vehicle Readiness Projects. http://www.afdc.energy.gov/uploads/publication/guide_ev_projects.pdf.

Frank, T., F. Neumann, F. Butler, P. Cocron, and J. Krems. 2011. Experiencing range in an electric vehicle: Understanding psychological barriers. Applied Psychology 61 (3): 368-391.

Friedman, M. 2014. "Fate of Bill to Allow Tesla to Restart Sales Could Hinge on Pro-Dealership Bill." NJ.com, July 5. http://www.nj.com/politics/index.ssf/2014/07/tesla_showrooms_previously_banned_may_return_to_new_jersey.html.

Gilbert, B. 2014. "Why Tesla Motors Can't Sell Cars in Most of the United States." Engadget, July 17. http://www.engadget.com/2014/07/17/tesla-motors-us-sales/.

GM Authority. 2010. "Chevrolet Announces Voltec 240V Home Charger For $490." http://gmauthority.com/blog/2010/10/chevrolet-annouces-voltec-240v-home-charger-for-490.

GSA (U.S. General Services Administration). 2011. "Electric Vehicle Pilot Program." http://www.gsa.gov/portal/content/281581?utm_source=OCM&utm_medium=print-radio&utm_term=electricvehicles&utm_campaign=shortcuts.

GSA. 2013. Federal Fleet Report. http://www.gsa.gov/portal/category/102859.

Hanson, S. 2013. "Frito-Lay Electric Vehicles: Why, How and What Is Next." Presentation to the Committee on Overcoming Barriers to Electric-Vehicle Deployment, Washington, DC, December 3.

Henry, J. 2012. "Average Car in the U.S. Now Over 10 Years Old, a Record." Forbes, January 20. http://www.forbes.com/sites/jimhenry/2012/01/20/average-car-in-the-u-s-now-over-10-years-old-a-record/.

Herbig, P.A., and H. Kramer. 1994. The effect of information overload on the innovation choice process: Innovation overload. Journal of Consumer Marketing 11(2): 45-54.

Hidary, J. 2012. "New Models of Mobility and EV Deployment." Presentation to the Committee on Overcoming Barriers to Electric-Vehicle Deployment, Washington, DC, December 18.

Ingram, A. 2013. "Many Consumers Still Unaware Of Electric Car Incentives." Green Car Reports, November 19. http://www.greencarreports.com/news/1088583_many-consumers-still-unaware-of-electric-car-incentives.

Ingram, A. 2014. "Electric Car 'Guest Drive' Service Launches in CA To Get 'Butts In Seats'." Green Car Reports, May 2. http://www.greencarreports.com/news/1091852_electric-car-guest-drive-service-launches-in-ca-to-get-butts-in-seats.

Kahle, L., and L. Chiagouris. 1997. Values, Lifestyles, and Psychographics. Mahwah, New Jersey: Lawrence Erlbaum Associates, Inc.

Kahneman, D. 2013. Thinking, Fast and Slow. New York: Farrar, Straus and Giroux.

Krause, R. M., S.R. Carley, B.W. Lanea, and J.D. Graham. 2013. Perception and reality: Public knowledge of plug-in electric vehicles in 21 U.S. cities. Energy Policy 63: 433-440.

LeBeau, P. 2012. "Americans Buying Fewer New Cars in Lifetime." CNBC, October 22. http://www.cnbc.com/id/49504504.

Leonard Barton, D., E.Wilson, and J. Doyle. 1995. Commercializing Technology: Understanding User Needs. In Business Marketing Strategy, edited by V.K. Rangan et al., 281-305. Chicago: Irwin.

Lim, M., H. Mak, and Y. Rong. 2014. "Toward Mass Adoption of Electric Vehicles: Impact of the Range and Resale Anxieties." http://papers.ssrn.com/sol3/papers.cfm?abstract_id=2236560.

Loveday, E. 2013a. "Consumer Reports Says Leasing is the Right Financial Choice for Today's Electric Vehicles." InsideEVs, June 28. http://insideevs.com/consumer-reports-says-leasing-is-the-right-financial-choice-for-todays-electric-vehicles/.

Loveday, E. 2013b. "Are Electric Vehicles Being Sold at a Loss? Kelley Blue Book Seems to Think So." InsideEVs, July 18. http://insideevs.com/are-electric-vehicles-being-sold-at-a-loss-kelley-blue-book-seems-to-think-so/.

Lutz, B. 2012. "The Real Story On GM's Volt Costs." Forbes, September 10. http://www.forbes.com/sites/boblutz/2012/09/10/the-real-story-on-gms-volt-costs/.

Mahajan, V., E. Muller, and F.M. Bass. 1990. New product diffusion models in marketing: A review and directions for research. The Journal of Marketing 54: 1-26.

McQuarrie, E. 2008. Customer Visits: Building a Better Market Focus. 3rd edition. Armonk, NY: Sharpe.

METI (Ministry of Economy, Trade, and Industry). 2010. Next Generation Vehicle Plan. http://www.meti.go.jp/english/press/data/pdf/N-G-V2.pdf.

Mohr, J., S. Sengupta, and S. Slater. 2010. Marketing of High-Technology Products and Innovations. 3rd edition. Upper Saddle River, NJ: Prentice Hall.

Mohr, J., S. Slater, and S. Sengupta. 2006. Foundations for Successful High-Technology Marketing. In Managing Technology and Innovation: An Introduction, edited by R. Verburg, R. J. Ortt, and W. M. Dicke, 84-105. London: Routledge Publishing.

Moore, G. 2014. Crossing the Chasm: Marketing and Selling Technology Products to Mainstream Customers. 3rd edition. New York: Harper Business.

Motavalli, J. 2013. "Ford Switches to AeroVironment for Chargers, Drops Best Buy." Plug-in Cars, May 10. http://www.plugincars.com/best-buy-out-aerovironment-fords-charging-partner-127183.html.

Nelson, G. 2014. "NADA Moves to Counter Franchise System Critics." Automotive News, June 16. http://www.autonews.com/article/20140616/RETAIL06/306169944/nada-moves-to-counter-franchise-system-critics.

NRC (National Research Council). 2013a. Transitions to Alternative Vehicles and Fuels. Washington, DC: The National Academies Press.

NRC. 2013b. Overcoming Barriers to Electric-Vehicle Deployment, Interim Report. Washington, DC: The National Academies Press.

Obama, B. 2011. "Federal Fleet Performance." Presidential memorandum. May 24. https://www.whitehouse.gov/the-press-office/2011/05/24/presidential-memorandum-federal-fleet-performance.

Parasuraman, A., and C. Colby. 2001. Techno-Ready Marketing: How and Why Customers Adopt Technology. New York: The Free Press.

Pike Research. 2012. Energy & Environment Consumer Survey Consumer Attitudes and Awareness about 13 Clean Energy Concepts: Published 1Q 2012. http://www.navigantresearch.com/wordpress/wp-content/uploads/2012/02/EECS-12-Pike-Research.pdf.

Plug-in America. 2013. "National Drive Electric Week 2013 Stats." https://driveelectricweek.org/stats.php?year=2013.

Quinland, R.M. 2013. Has the traditional automobile franchise system run out of gas? The Franchise Lawyer 16(3). http://www.americanbar.org/publications/franchise_lawyer/2013/summer_2013/has_traditional_automobile_franchise_system_run_out_gas.html.

Rai, V., and V. Nath. 2014. "How the Interaction of Supply and Demand Shapes Patterns of New Technology Adoption: Plug-In Electric Vehicles in California." Paper presented at the International Association for Energy Economics, New York, NY, June 16.

Rayport, J., and D. Leonard Barton. 1997. Spark innovation through empathic design. Harvard Business Review 75(6): 102-113.

Richard, M.G. 2010. "GE to Purchase 25,000 Electric Cars by 2015!" TreeHugger, November 11. http://www.treehugger.com/cars/ge-to-purchase-25000-electric-cars-by-2015.html.

Rogers, E. 2003. The Diffusion of Innovations. 5th edition. New York: The Free Press.

Scheibehenne, B., R. Greifeneder, and P.M. Todd. 2010. Can there ever be too many options? A meta-analytic review of choice overload. Journal of Consumer Research 37(3): 409-425.

Schwartz, B. 2005. Paradox of Choice: Why More Is Less. New York: Harper Perennial.

Seybold, P. 2001. Get inside the lives of your customers. Harvard Business Review 79(5): 81-89.

Shaheen, S., and A. Cohen. 2013. "Transportation Sustainability Research Center: North American Carsharing 2012-2013 Outlook." 2013 Carsharing Association Conference. Toronto, Canada, September 16-17. http://www.slideshare.net/AlanWoodland/dr-susan-shaheen-north-american-carsharing-20122013-outlook.

Shepard, S., and J. Gartner. 2014. Electric Vehicle Geographic Forecasts. Navigant Research. Boulder, CO.

Sivak, M. 2014. Has Motorization in the U.S. Peaked? Part 4: Households without a Light-Duty Vehicle. Report No. 2014-5. University of Michigan Transportation Research Institute, Ann Arbor, MI. http://deepblue.lib.umich.edu/bitstream/handle/2027.42/102535/102988.pdf.

Solomon, M. 2014. Consumer Behavior: Buying, Having, and Being. 11th edition. Boston: Prentice Hall.

Strategic Vision. 2013. New Vehicle Experience Studies of Vehicle Registrants. San Diego, CA.

Strategic Vision. 2014. New Vehicle Experience Studies of Vehicle Registrants. San Diego, CA.

Sultan, F., J. U. Farley, and D. R. Lehmann. 1990. A meta-analysis of applications of diffusion models. Journal of Marketing Research 27(1): 70-77.

Tal, G., M.A. Nicholas, J. Woodjack, and D. Scrivano. 2013. Who Is Buying Electric Cars in California? Exploring Household and Vehicle Fleet Characteristics of New Plug-in Vehicle Owners. Research Report UCD- ITS-RR-13-02. Institute of Transportation Studies, University of California, Davis.

Turrentine, T.S., D.M. Garas, A.H. Lentz, and J.F. Woodjack. 2011. The UC Davis MINI E-Consumer Study. UCD-ITS-RR-11-05. Institute of Transportation Studies, University of California, Davis.

UC Davis. 2014. "Plug-in Hybrid and Electric Vehicle Research Center of the Institute of Transportation Studies." PEV Market Briefing, May 2014. http://policyinstitute.ucdavis.edu/files/PEV-Market-Briefing-Turrentine-28May2014.pdf.

U.S. Census Bureau. 2012. "2012 Economic Census, 2012 Economic Census of Island Areas, and 2012 Nonemployer Statistics." Industry Statistics Portal NAICS 4471. http://www.census.gov/econ/isp/.

U.S. Census Bureau. 2013a. Income, Poverty, and Health Insurance Coverage in the United States: 2012. https://www.census.gov/prod/2013pubs/p60-245.pdf.

U.S. Census Bureau. 2013b. "Physical Housing Characteristics for Occupied Housing Units: 2009-2013 American Community Survey 5-year Estimates." American FactFinder. http://factfinder.census.gov/bkmk/table/1.0/en/ACS/13_5YR/S2504/0100000US. Accessed January 22, 2015.

Voelcker, J. 2013a. "Next Chevy Volt $10,000 Cheaper to Build, Profitable for GM?" Green Car Reports, May 2. http://www.greencarreports.com/news/1083890_next-chevy-volt-10000-cheaper-to-build-profitable-for-gm.

Voelcker, J. 2013b. "Why Some Dealers Are Inept at Selling Plug-In Electric Cars." Green Car Reports, December 12. http://www.greencarreports.com/news/1089055_why-some-dealers-are-inept-at-selling-plug-in-electric-cars.

Voelcker, J. 2014. "Many Car Dealers Don't Want to Sell Electric Cars: Here's Why." Green Car Reports, February 14. http://www.greencarreports.com/news/1090281_many-car-dealers-dont-want-to-sell-electric-cars-heres-why.

Wells, W. 2011. Life Style and Psychographics. Chicago: Marketing Classics Press.

Westervelt, A. 2012. "Electric Vehicles Lead Pack in Greening Corporate Fleets." Green Biz, March 30. http://www.greenbiz.com/blog/2012/03/30/electric-vehicles-corporate-fleet.

Wolski, C. 2013. "What Is the Future of Electric Vehicles in Fleet?" Green Fleet, September. http://www.greenfleetmagazine.com/article/51755/what-is-the-future-of-electric-vehicles-in-fleet/p/2.

Zmud, J., V. Barabba, M. Bradley, J.R. Kuzmyak, M. Zmud, and D. Orrell. 2014. Strategic Issues Facing Transportation: Volume 6, The Effects of Socio-Demographics on Future Travel Demand. NCHRP Report 750. Transportation Research Board, Washington, DC. http://onlinepubs.trb.org/onlinepubs/nchrp/nchrp_rpt_750-v6.pdf.

4

Government Support for Deployment of Plug-in Electric Vehicles

The successful deployment of plug-in electric vehicles (PEVs) involves many entities and will require the resolution of many complex issues. The present report focuses on individual strategies for overcoming barriers related to purchasing and charging PEVs, and this chapter specifically explores how federal, state, and local governments and their various administrative arms can be more supportive and implement policies to sustain beneficial strategies for PEV deployment. Although electric utilities can also provide institutional support for PEV deployment, they and their associated policies are discussed in Chapter 6. Where opportunities exist to improve the viability of PEVs but no single institution is clearly positioned to capitalize on the opportunity, the committee highlights possible partnerships that might fill these voids. The committee's findings and recommendations are provided throughout this chapter.

FEDERAL GOVERNMENT RESEARCH FUNDING TO SUPPORT DEPLOYMENT OF PLUG-IN ELECTRIC VEHICLES

Funding research is one of the most important ways the federal government can lower barriers to PEV deployment. Research is needed in two areas in particular. As discussed in Chapter 2, the first is basic science and engineering research to lower the cost and improve the energy density and other performance characteristics of batteries. The second critical area concerns PEV deployment, especially the role of infrastructure in spurring vehicle sales and increasing electric vehicle miles traveled (eVMT). Fundamental and applied science and engineering research for vehicle energy storage is being undertaken by vehicle manufacturers and in the laboratories of the U.S. Department of Energy (DOE), the Department of Defense (DOD), and academic institutions. Research into the deployment of PEV infrastructure and markets is much less developed. Both areas are discussed below.

Engineering Research and Development of Battery Science

As discussed in Chapter 2, the battery is the most costly component of PEVs and represents the majority of the cost differential between PEVs and internal-combustion engine (ICE) vehicles. Battery cost will need to decrease substantially to allow PEVs to become cost competitive with ICE vehicles (see Chapters 2 and 7). Thus, the current goal of battery research and development is to increase the energy density of PEV batteries and to lower their cost. The improved battery technology can then be used to lower vehicle cost, increase vehicle range, or both, and those improvements would likely lead to increased PEV deployment.

As in many areas of fundamental research and development, the federal government has an important role to play. Although basic science and engineering research is funded by both government and the private sector, the government role is to fund long-term, exploratory research that has the potential for positive national impact. Stable funding for exploratory research allows investments in research facilities and human capital that are necessary for the research to bear fruit. The federal government has directly supported battery research and development for electric vehicles since 1976 (Electric and Hybrid Vehicle Research, Development, and Demonstration Act 1976, Pub. L. 94-413). Past investment in research and development contributed to the development of the NiMH batteries used in early hybrid electric vehicles (HEVs) and to the lithium-ion battery technology used in the Chevrolet Volt (DOE 2008).

The largest funder of energy storage research in the federal government is DOE, followed by DOD. From 2009 to 2012, across all areas of the federal government, investment in energy-storage research, development, and technology deployment totaled $1.3 billion, which includes batteries for all applications, not only vehicles (GAO 2012a). In Fiscal Year (FY) 2013, the DOE Vehicle Technology Office funded $88 million for battery research and development focused on vehicle applications (DOE 2014a). Much of the funding is for grants or cooperative research agreements with government, industry, or university laboratories, but a growing proportion is also funding loan guarantees to deploy new technologies. Worthy DOE goals for battery storage improvements include halving the size and weight of PEV batteries and reducing the production costs to one quarter of its 2012 value by 2022 (DOE 2013a). Recently, DOE has initiated and supported several collaborative research programs with even more ambitious goals to accelerate basic and applied research, development, and deployment. They include Energy Frontier Research Centers, several Advanced Research Projects Agency-Energy (ARPA-E) programs in energy storage, and the Battery and Energy Storage Hub, which is funded at up to $25 million per year for 5 years and aims to increase battery energy density five times and reduce cost by 80 percent (DOE 2013b).

Finding: Investment in battery research is critical for producing lower cost, higher performing batteries that give PEVs the range consumers expect from ICE vehicles.

Recommendation: The federal government should continue to sponsor fundamental and applied research to facilitate and expedite the development of lower cost, higher performing vehicle batteries. Stable funding is critical and should focus on improving energy density and addressing durability and safety.

Research on Deployment of Plug-in Electric Vehicles

In contrast to the substantial investment in battery research and development, research on PEV deployment is much less advanced. A critical research need is understanding the relationship between PEV deployment and infrastructure deployment. Supporting that research is an appropriate role for the federal government given that it might be motivated to deploy infrastructure if by doing so it encourages PEV deployment and increases eVMT.

The primary DOE effort to understand PEV vehicle and infrastructure deployment is the EV Project, an infrastructure deployment and evaluation program managed by the Idaho National Laboratory (INL) in partnership with ECOtality. Around the time of the most recent wave of PEVs in 2009, DOE awarded in 2009 a $99.8 million grant for deployment of charging infrastructure in private residences and in public areas in 20 of the target launch markets of the Nissan Leaf and the Chevrolet Volt, including San Francisco, Seattle, San Diego, Los Angeles, Portland, and Nashville. The program has grown with an additional $15 million grant from DOE and partner matches from the vehicle manufacturers and charging providers to a total of $230 million (ECOtality 2013; INL 2014a). When it concluded collecting data in December 2013, over 8,200 vehicles were participating and over 8,200 residential chargers, 3,500 public AC level 2 chargers, and 107 DC fast chargers had been installed (Smart and White 2014; INL 2014b).

The EV Project included data collection on where and when the vehicles in the project charged so that DOE could learn more about how drivers were using the vehicles and the associated charging infrastructure. Thus, the data provided important information about early adopters of PEVs in large metropolitan areas, including location of charging, eVMT, impacts on utilities, impact of workplace charging, and regional variations in charging behavior. Because privacy is an important consideration in the United States, there were clearly limitations on the tracking data that could be shared with researchers. Data collection ended as of December 2013, but data analysis continues.

Finding: Research is critically needed in understanding the relationship between infrastructure deployment and PEV adoption and use.

Recommendation: The federal government should fund research to understand the role of public charging infrastructure (as compared with home and workplace charging) in encouraging PEV adoption and use.

Recommendation: A new research protocol should be designed that would facilitate access to raw charging data to relevant stakeholders within the confines of privacy laws.

INSTITUTIONAL SUPPORT FOR PROMOTING PLUG-IN ELECTRIC VEHICLE READINESS

The concept of *PEV readiness* refers to an entire ecosystem of automotive technology, including its supporting infrastructure, regulations, financial incentives, consumer information, and public policies, programs, and plans that can make PEVs a viable choice for drivers. Several tools have been created to assess whether a given organization, community, state, or even country has in place the essential elements to be considered PEV ready. Examples of assessment tools include the Rocky Mountain Institute's *Project Get Ready* (Rocky Mountain Institute 2014), DOE's *Plug-in Electric Vehicle Readiness Scorecard* (DOE 2014a), Michigan Clean Energy Coalition's *Plug-in Ready Michigan* (Michigan Clean Energy Coalition 2011), California PEV Collaborative's PEV *Readiness Toolkit* (CAL PEV 2012a), and the Center for Climate and Energy Solutions' *State DOT PEV Action Tool* (C2ES 2014). Furthermore, $8.5 million has been provided through the DOE Clean Cities program to 16 projects across 24 states to assess PEV readiness and develop specific plans to enable the communities to become PEV ready (DOE 2014a). Table 4-1 indicates the many common factors that constitute PEV readiness and the different institutions or organizations that might have a role to play.

State governments will be particularly important actors in supporting PEV deployment. Most supportive PEV actions at the state level can be carried out by various administrative agencies, including environmental and clean air agencies, utility commissions, departments of energy, transportation agencies, licensing and inspection agencies, general services agencies, and workforce training or education agencies. In the committee's interim report, the committee noted several areas where the federal government could play a convening role to coordinate state and local government activities in support of the emerging PEV sector (NRC 2013, pp. 2, 4, 52).

TABLE 4-1 Factors Determining PEV Readiness and Organizations Involved

Readiness Feature	Federal Government	State Government	Municipal Government	Electric Utility	Private Industry
Permit streamlining	—	Environmental and archeological	Building and electrical codes	—	—
Utility regulatory policies	—	PUC regulation of cost recovery and retail markets	Muni-owned cost recovery policies	—	—
Building code requirements	Model ordinances	Model state ordinances	Local ordinances	—	PEV-ready buildings
Infrastructure deployment plans	DOE funding, assistance, and dissemination	Interregional and interstate plans	Regional and metropolitan area plans	Distribution network and capacity	Strategic investment plans and sites
Land use and uniform signage	Federal regulations	State regulations and policies	Comprehensive plans and zoning	—	—
Electricity pricing policies	NIST metering and pricing standards	State laws and PUC rate regulation	Muni-owned policies and technology	Smart grid and metering technologies	EVSE pricing strategies
Training personnel	—	Workforce training and permits	First-responder safe practices	—	Skilled trades
Vehicle financial incentives	PEV subsidies	Rebates, tax exemptions from registration, tolls	Utility taxes, parking fees	Rebates	Equity investments, financing
Infrastructure financial incentives	Equipment subsidies	Equipment subsidies	Equipment subsidies, land gifts	Cost sharing in any upgrades, equipment subsidies	Workplace and fleet charging
Energy policies	Clean energy programs	Zero-emission-vehicle standards	TOU or special PEV rates	TOU or special PEV rates	Green power programs
Dealership franchise laws	—	State laws and regulations	—	—	Vehicle manufacturers' policies and practices
Environmental policies	EPA regulations	Clean air laws and regulations	Carbon reduction plans	Clean power generation	—
Procurement policies and goals	GSA regulations and policies	State purchasing	Purchasing cooperatives and bulk orders	—	Bulk purchase discounts
Business policies and permissible models	Research and demonstration projects	State-backed financing assistance	Municipal-owned infrastructure	Own or operate EVSE	Innovative financing

NOTE: DOE, U.S. Department of Energy; EPA, U.S. Environmental Protection Agency; EVSE, electric vehicle supply equipment; GSA, General Services Administration; NIST, National Institute of Standards and Technology; PEV, plug-in electric vehicle; PUC, public utility commission; TOU, time of use.

TRANSPORTATION TAXATION AND FINANCING ISSUES RELATED TO PLUG-IN ELECTRIC VEHICLES

One potential barrier for PEV adoption that is solely within the government's direct control is taxation of PEVs,[1] in particular, taxation for the purpose of recovering the cost of maintaining, repairing, and improving the roadways. As described below, the paradigm for roadway taxation in the United States has depended on motor fuel taxes, which are indirect user fees. The advent of PEVs poses a dilemma for public officials responsible for transportation-tax policy because battery electric vehicles (BEVs) use no gasoline and plug-in hybrid electric vehicles (PHEVs) use much less than ICE vehicles.[2] To further complicate matters, there appears to be widespread misunderstanding about the extent to which PEVs currently pay transportation taxes and the resulting fiscal impacts to transportation budgets both now and into the future. This section explores the issue in depth, attempts to bring more clarity to current tax policy and impacts, and makes recommendations for how transportation-tax policy might be harmonized with a transportation innovation policy for PEVs.

Current State of Transportation Taxation

Motor fuel taxes have been the most important single source of revenue for funding highways for nearly a century and have also been an important source of transit funding since the 1980s (TRB 2006, pp. 24-36). The state of Oregon instituted the nation's first per-gallon tax on gasoline in 1919 (ODOT 2007). Within 10 years after that, every state had enacted a fuel tax. The federal government did not enact a fuel tax until 1932 and did not dedicate the tax to transportation

[1] Chapter 7 addresses the issue of tax incentives; this chapter discusses tax *dis*incentives.

[2] The amount of gasoline used by a PHEV depends on the all-electric range and the frequency with which the vehicle is charged.

projects until 1956 (FHWA 1997, Chapter IV). At the time of their introduction, fuel taxes were viewed as the most economical method of collecting a fee for roadway construction and maintenance from those who directly benefited: motor-vehicle operators. However, the share of highway spending covered by fuel-tax revenues has been declining. In 2012, fuel taxes accounted for 59 percent of all federal, state, and local highway-user revenues (fuel taxes, fees, and tolls) used for highways and 28 percent of total government disbursements for highways (FHWA 2014, Table HF-10).

For most of the past century, the fuel tax has been viewed as a reasonably fair and reliable tax revenue to fund transportation. The fuel economy of most vehicles remained fairly consistent across different models (NSTIFC 2009) as there were no strong incentives (such as increasing gasoline prices or stricter government regulation) to improve fuel economy. However, the 1973 Yom Kippur War and resulting oil Arab embargo served as the marker for the U.S. policy shift to reduce the nation's petroleum dependence by improving vehicle fuel economy. In later years, the federal government enacted Corporate Average Fuel Economy (CAFE) regulations, which essentially mandated improved fuel economy in passenger vehicles (see Figure 4-1).

Both the federal government (see Figure 4-2) and the states rely heavily on motor-fuel tax revenue, which includes taxes on gasoline and diesel, to maintain the transportation system. At the federal level and in the vast majority of states, fuel taxes are based on a flat cents-per-gallon tax levied on motor fuel; the extent of reliance on the fuel taxes varies from state to state (Rall 2013). For example, gasoline taxes range from $0.08 per gallon in Alaska to $0.53 per gallon in California (the nationwide average is $0.31 per gallon) (Rall 2013). Of all government tax and fee revenues used for highways in 2012, 20 percent came from the federal government, 49 percent from state governments, and 31 percent from local governments (FHWA 2014, Table HF-10).

Fuel consumption depends on both the number of miles driven and the fuel economy of the vehicle fleet. Therefore, any decrease in the number of miles driven or increase in the fuel economy of the vehicle fleet will result in less tax revenue generated for a cost-per-gallon tax. One of those factors can offset the other and moderate the negative effect on the revenue stream. For example, the fuel economy of the light-duty vehicle fleet has been increasing since 2005 (EPA 2013). From 2005 to 2007, light-duty vehicle miles traveled (VMT) also increased, which helped mask the negative effect on the revenue stream of improving fleet fuel economy. However, VMT and fuel consumption both declined with the recession in 2007 and 2008 and have remained flat since then (Figure 4-3). Without the revenue-bolstering effect from increasing VMT, transportation budgeters and policy makers have become acutely aware of how rising fleet vehicle economy affects transportation fund balances.

Federal and State Concerns

With the recent increases in federal CAFE standards,[3] the flattening of VMT, and political opposition to raising the tax rate itself, federal and state officials are increasingly concerned with the potential effects of high-mpg vehicles on their transportation budgets. The poster child for their worries is the BEV, which uses no gasoline and whose drivers therefore pay no fuel tax.

A recent survey of 50 state departments of transportation (DOTs) reflected the strong sentiment that PEVs threaten loss of revenue for transportation. The majority of state DOTs responded that they would support federally led field tests of mileage fees for PEVs to improve the equity and sustainability of Highway Trust Fund revenues (GAO 2012b, p. 45).

[3] 49 CFR Parts 523, 531, 533 et al., 2017 and Later Model Year Light-Duty Vehicle Greenhouse Gas Emissions and Corporate Average Fuel Economy Standards; Final Rule.

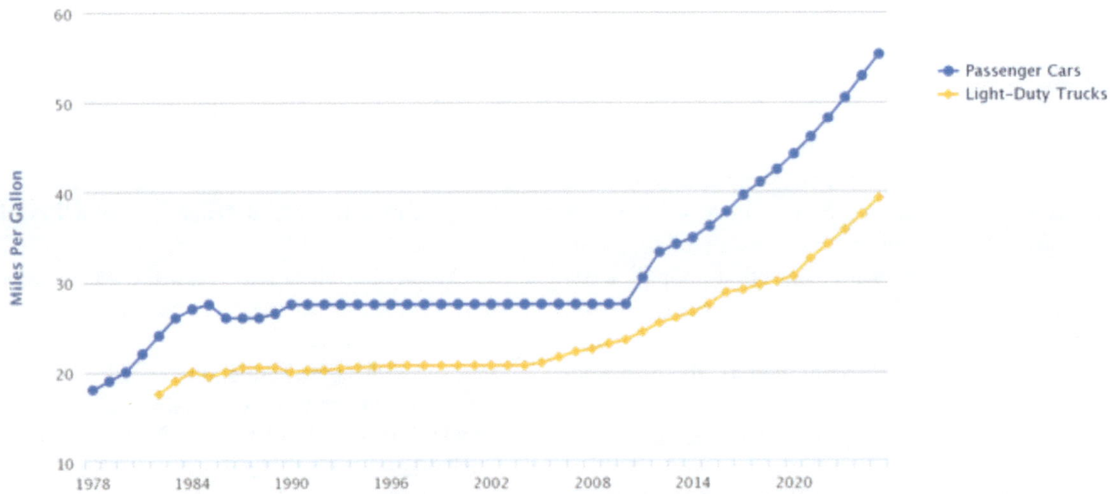

FIGURE 4-1 Corporate Average Fuel Economy requirements by year. SOURCE: DOE (2013c).

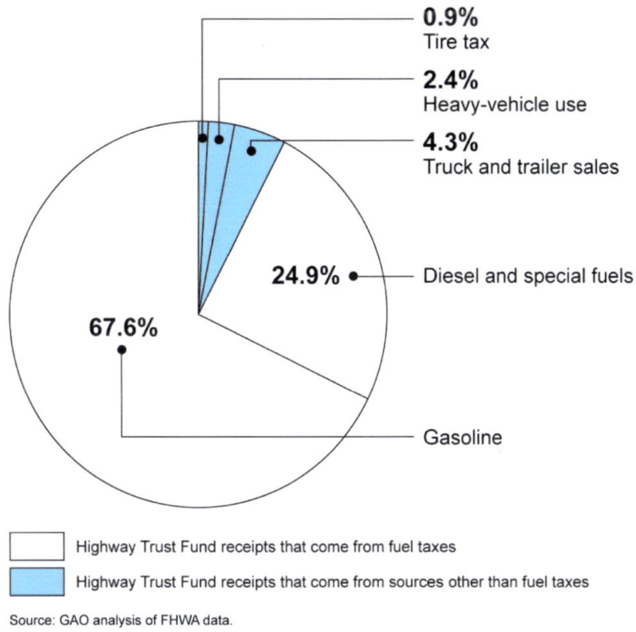

FIGURE 4-2 Sources of revenue for the federal Highway Trust Fund, FY 2010. These revenue sources *exclude* transfers from the general fund because those are not considered *revenues* in the federal nomenclature. SOURCE: GAO (2012b, p. 6).

A common refrain is that "PEVs pay nothing to use the highways" because they use little if any gasoline (Battaglia 2013). That is not, however, the case. At the federal level, the highway trust fund has relied on transfers of general tax revenues to maintain sufficient balances to meet its transportation funding obligations (GAO 2011). Therefore, all U.S. taxpayers—including PEV drivers—are paying for the federal transportation system from their general tax payments, in addition to the 18.4 cent per gallon federal gasoline tax.

That misunderstanding is even more acute at the state level, where many states and local governments levy a myriad of taxes and fees that are dedicated to transportation, including roadway funding.[4] Specifically, most local transportation funding comes from property taxes, general fund appropriations, and fares for mass transit; at the state level, motor fuel taxes are significant, but motor vehicle taxes, fees, and other revenue, such as sales taxes, play important roles. Washington State recently estimated that, on average, BEV drivers pay $210 per year in transportation-related state and local taxes and fees even though they pay no fuel taxes (WSDOT 2013).[5] That equates to 44 percent of what is paid by the average gasoline-powered passenger vehicle in that state. Figure 4-4 compares transportation-related taxes paid by Washington state drivers of different classes of vehicles.

[4] For a breakdown of transportation funding sources at the federal, state and local levels, see http://www.transportation-finance.org/funding_financing/funding/.

[5] Calculations are based on the 11,489 miles per year driven, on average, by drivers residing in the greater Seattle metro area.

The committee recognizes that PEVs and current transportation tax policies raise the following important questions:

- Is the difference in transportation taxes collected from PEVs and ICE vehicles significant in the context of federal or state transportation budgets, either now or in the near future?
- Even if the amount of unrealized revenue is negligible, do PEVs raise issues of fairness in the user-pays principle underlying the U.S. transportation tax system that has been in place for almost a century?
- To remedy the issues inherent in the first two questions, should PEVs be a focus for new methods of taxation, considering that the unrealized revenue from high-mpg vehicles will dwarf that of PEVs?
- Are there other intervening policy considerations that might trump the general transportation tax paradigm of user pays, at least for a period of time?

Finding: It is not true that PEV drivers pay nothing for the maintenance and use of the transportation system given various transportation-related taxes and fees that must be paid by all vehicle drivers. It is true that BEVs pay no federal or state gasoline taxes, and it is also true that PHEVs, such as the Chevrolet Volt, might pay proportionately very little in gasoline taxes.

Recommendation: Governments (federal, state, and local) should fully and fairly disclose all transportation-related tax-

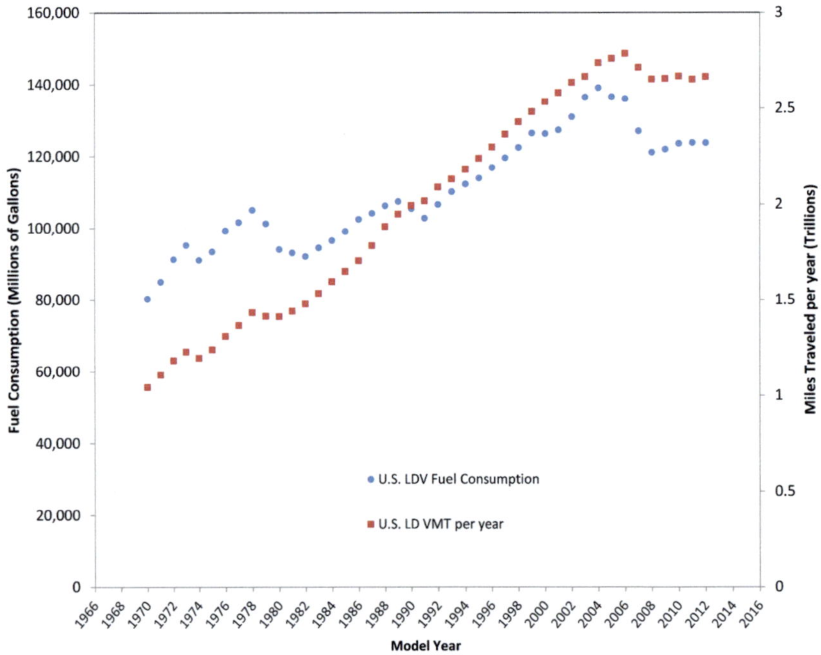

FIGURE 4-3 U.S. annual light-duty fuel consumption and VMT. NOTE: VMT, vehicle miles traveled; LDV, light-duty vehicle. SOURCE: DOE (2014b).

es and fees currently paid by all vehicles, including average passenger vehicles, alternative-fuel vehicles (such as compressed natural gas), HEVs, PHEVs, and BEVs. Providing that information to elected officials and the public will give them an accurate baseline against which policy discussions and choices can be made.

Impacts on Transportation Budgets

As noted, the first policy question is whether, from a fiscal viewpoint, the lack of fuel taxes paid by PEVs is having a negative effect on federal or state transportation budgets, either now or within the next 10 years. At the federal level, estimates can be made of the unrealized fuel tax revenues from PEVs; the results are shown in Table 4-2. On the basis of the number of PEVs sold through 2013, an additional $14 million annually could be generated for the federal Highway Trust Fund if each PEV was required to pay $96 per year, the same amount paid by a driver of a 22 mpg gasoline-powered sedan. To put that amount in context, the federal Highway Trust Fund collects fuel-tax revenues of about $33 billion each year (CBO 2013).

PEV industry analysts have also examined the impact of PEVs on transportation budgets. The California PEV Collaborative—a public-private consortium of governments, private businesses, vehicle manufacturers, and nongovernment organizations allied to promote PEVs—recently found that if the Obama administration goal of putting 1 million PEVs on the road by 2015 were met with BEVs, the resulting unrealized revenue from motor fuel taxes would be less than 0.5 percent of the total projected revenue shortfall for the federal Highway Trust Fund (CAL PEV 2012b).

Finding: For the next few years, the fiscal impact of not collecting a fuel tax from PEVs is negligible.

Fairness and Equity in Transportation Taxes

The second policy question is whether PEV drivers who pay little or no fuel taxes raise issues of fairness, given the strong user-fee paradigm for funding the expenses of the highway infrastructure in the United States. Even though the government would only derive an extremely small share of revenue by taxing PEVs, the sentiment among elected officials and the general public remains that PEV drivers should be paying the fuel tax (or its equivalent) as their fair share for maintaining and improving the roadways on which they drive. Although its study did not focus on equity issues related to taxation of PEVs, TRB (2011) did identify strongly held notions of fairness and equity that are inherent in the transportation tax system and that are important for public policy making; they are summarized in Table 4-3.

The fairness issue in the tax treatment of PEVs appears to be more acute at the state and local levels, where many elected officials are actively considering fuel-tax increases to reduce the backlog of roadway maintenance and improve-

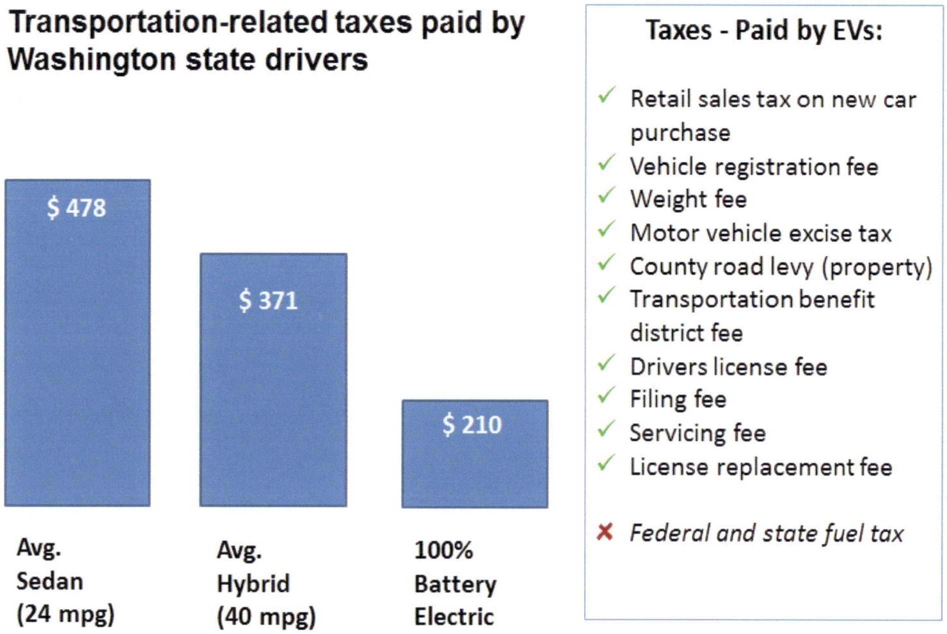

FIGURE 4-4 Annual transportation-related taxes paid by Washington state drivers. SOURCE: WSDOT (2013).

ment projects. As noted by the TRB (2011, p. 103), for politicians and other decision makers, one of the first hurdles to overcome in embarking on a new transportation initiative—which will require financing, perhaps through an increase in the fuel tax—is to gain public support. Decision makers go to great lengths to ensure that the burdens (taxes) and benefits (capital projects) are allocated in ways that are perceived as fair. It is in trying to rally public support for tax increases that some politicians have sought to remedy the perceived unfairness concerning unrealized revenue from PEV drivers (Vekshin 2013). Washington, Virginia, North Carolina, South Carolina, New Jersey, Indiana, Arizona, Michigan, Oregon, and Texas have all considered or, in some cases, enacted legislation that imposes a fee or tax on PEVs. Many of the efforts were undertaken as part of, or coincident with, proposals to increase the fuel tax on all motorists.

Finding: Perceptions of fairness and equity are important factors to consider in PEV tax policies, even though the actual revenue impact of PEV taxation is negligible in the short run and likely to remain minimal over the next decade.

Government Responses to Plug-in Electric Vehicles and High-Mileage Vehicles

The third policy question raised is the extent to which PEVs should be a specific focus for new methods of taxation, considering the much larger impact other high-mileage vehicles will have on transportation funding levels, particularly once the 2025 CAFE standards (54.5 mpg) take effect. The fuel economy of the entire light-duty passenger vehicle fleet is increasing and will continue to increase in the coming decades largely due to federal CAFE standards (see Figure 4-1). The Congressional Budget Office (CBO) estimated that the new CAFE standards would gradually lower federal gasoline-tax revenues, eventually causing them to fall by 21 percent. The CBO analysis demonstrated that from 2012 through 2022, which is before the most stringent CAFE standards take effect, there will be a $57 billion drop in revenues (CBO 2012).

In addition to federal consideration of the impacts of high-mileage vehicles, many states are now actively exploring potential solutions to the forecasted revenue shortfalls (see Figure 4-5). At least one state (Washington) has forecast the potential transportation-revenue shortfalls attributable to improving fuel economy and to alternative-fuel vehicles, such as PEVs, and found that the potential drop in revenues ranges from 10 to 28 percent over the next 25 years (WSTC 2014).

Both federal and state policy makers and the public are becoming increasingly aware of the impact that high-mileage and alternative-fuel vehicles will have on roadway funding (Weissmann 2012). The Texas Transportation Institute recently convened several focus groups to better understand public sentiment. Participants strongly preferred mileage fees for vehicles that might only pay state vehicle registration and title fees for their road use (GAO 2012b).

TABLE 4-2 Comparison of Unrealized Revenue from Battery Electric Vehicles and Plug-in Hybrid Electric Vehicles

Vehicle Type	U.S. Total 2013[a]	Average Annual VMT	Fuel Economy (MPG or MPGe)	Annual Gallons Consumed	Federal Gas Tax Rate	Annual Unrealized Revenue
Avg. Sedan[b]	—	11,489	22	522 gal	$0.184	$96 per vehicle[d]
BEV	72,028	11,489	—	—	$0.184	$6.9 million
PHEV[c]	95,589	11,489	98	117 gal	$0.184	$7.1 million

[a] Electric Drive Transportation Association Sales Dashboard, Totals from December 2010 to December 2013.
[b] The data comprising the base case are adapted from GAO (2012b, p. 9).
[c] Because PHEV models vary widely, the Chevrolet Volt was used as the reference case as it has the longest all-electric range of the PHEVs on the market.
[d] This estimate is the baseline annual gasoline tax paid per vehicle, not annual unrealized revenue.
NOTE: BEV, battery electric vehicle; MPG, miles per gallon; MPGe, miles per gallon gasoline equivalent; PHEV, plug-in hybrid electric vehicle; VMT, vehicle miles traveled.

TABLE 4-3 Types of Equity and Examples in the Transportation Tax System

Type of Equity	Simple Definition	Transportation Example
Benefits received	I get what I pay for	People who use a facility the most pay the most.
Ability to pay	I pay more because I have more money	A project is financed through a progressive tax that is disproportionately paid by higher income people.
Return to source	We get back what we put in	Transit investment in each county is matched to that county's share of metropolitan tax revenues used for transit.
Costs imposed	I pay for the burden I impose on others	Extra expense required to provide express bus service for suburb-to-city commuters is recovered by charging fares for this service.
Process (or participation)	I had a voice when the decision was made	Public outreach regarding proposed new high-occupancy-toll lanes provides transparent information and seeks to involve all affected parties in public hearings and workshops.

SOURCE: TRB (2011, p. 41).

Whether the concern is limited to PEVs or more broadly centered on high-mileage vehicles, states are beginning to take action. Several states have enacted special taxes on PEVs or are considering how to tax them. Other states are exploring new transportation-tax methods to address not only PEVs but all high-mileage vehicles (see Figure 4-6).

Two congressionally chartered transportation funding and financing commissions—the National Surface Transportation Policy and Revenue Study Commission and the National Surface Transportation Infrastructure Financing Commission—have independently called for a transition from the current fuel-tax system to a mileage-based fee system to fund the nation's highway infrastructure (NSTPRSC 2007, pp. 51-54; NSTIFC 2009, p. 7). A recent report by the Government Accountability Office (GAO) examined the feasibility of mileage fees and recommended a federally sponsored pilot program to evaluate the viability, costs, and benefits of mileage fee systems, particularly for commercial trucks and PEVs (GAO 2012b). GAO (2012b, p. 45) found that two-thirds of state DOTs (34 of 51, including the District of Columbia) reported that they would support federally led field tests of mileage-based fees for PEVs; none reported that they would be opposed to such tests for PEVs. The Road Usage Fee Pilot Program Act of 2013 was introduced in the U.S. House of Representatives to authorize, fund, and partner with states to conduct VMT pilot projects across the nation.

Separate from the federal government efforts, over 20 states are actively studying, testing, or, in the case of Oregon, implementing some version of a mileage-based fee, also known as road usage charges or VMT fees or simply taxes (D'Artagnan Consulting 2012). The fundamental concept is that drivers would be assessed a cents-per-mile tax for every mile that is driven within the taxing jurisdiction (region, state, or nation), regardless of the vehicle type, fuel source, or engine technology.

Recommendation: In jurisdictions that do impose special taxes, fees, or surcharges on PEVs as a means of requiring contribution to roadway upkeep, governments should ensure that such taxes are proportionate to actual usage, just as current motor fuel taxes are proportionate to usage.

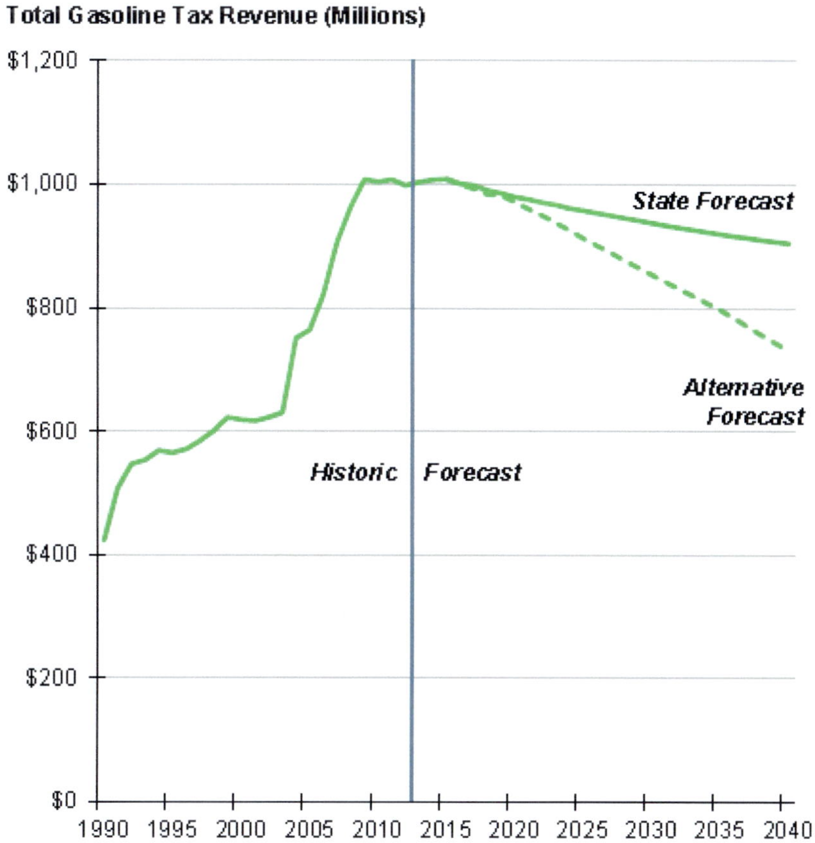

FIGURE 4-5 Historic and forecast gasoline-tax revenue for Washington state, FY 1990 to FY 2040. SOURCE: WSTC (2014, p. 5).

Intervening Policy Considerations in the Taxation of Plug-in Electric Vehicles

The last policy issue examined is whether other intervening policy considerations might trump the general transportation-tax paradigm of user pays, at least until PEVs have reached some level of market penetration. U.S. tax policy has a long and successful track record of encouraging innovation (Reuters 2013). There are many examples in the current U.S. tax code (and state tax codes) where taxes are exempted, credited, or rebated to promote the development or proliferation of services, assets, or activities deemed to provide a public benefit, such as dependent-care tax benefits and research and development or manufacturing tax credits. That tax forbearance acts as the public's investment in the societal good produced.

Most tax incentives are limited in scope, duration, or amount, so as to target more carefully the specific activity to be encouraged and to limit the public's subsidization (or investment). The current federal $7,500 tax credit for PEVs is a good example of a narrowly targeted federal subsidy (IRS 2009). As currently enacted, the amount of the credit increases on the basis of the capacity of the PEV battery because the battery is the most expensive component unique to PEVs and most in need of technological breakthrough. The tax credit is also limited in the amount available per taxpayer ($7,500) and limited in duration (credit is phased out after the manufacturer reaches vehicle sales of 200,000).

In contrast, there is no intentional or targeted tax incentive to encourage PEVs to drive on public roadways.[6] Instead, the government's pro-PEV scheme consists of tax credits, rebates, fee reductions, and exemptions for the purchase and ownership of the PEV—but not for its use of public roadways. The fact that PEVs do not pay the fuel tax or a similar road usage tax stands apart from the vast majority of tax policies that are transparent, legislatively granted, and targeted in scope, quantity, or duration.

To the extent policy makers wish to continue providing PEV drivers with the financial benefit of not paying the fuel tax (or alternative road user charge), serious consideration should be given to explicitly and intentionally adopting such a policy in the same manner as other tax incentives. Although it might initially seem odd to enact a law or regulation that specifically exempts an activity (PEV driving) that is already

---

[6] One could argue that allowing PEVs to drive in the high-occupancy-vehicle lane is an incentive to drive, as opposed to an incentive to own, and that the resulting loss of occupancy in the lane for other vehicles represents a public "investment."

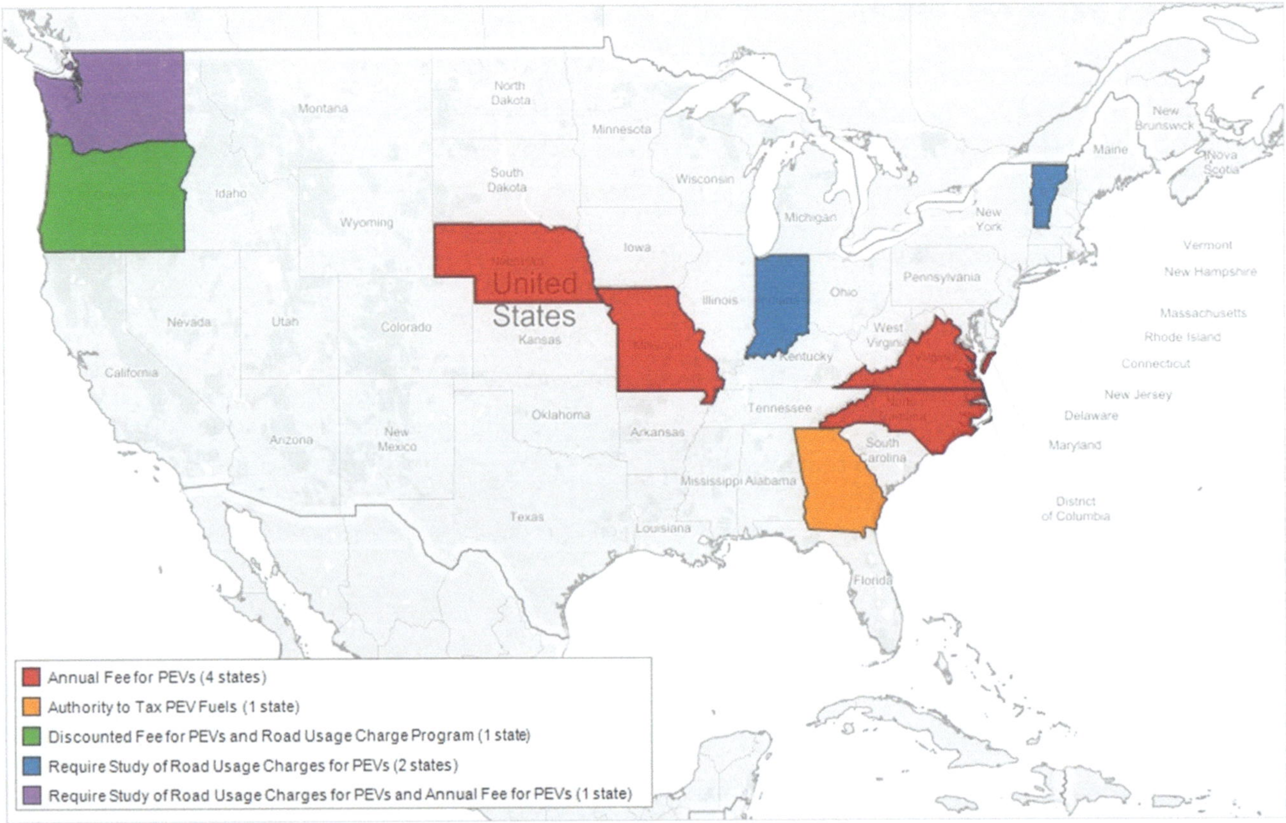

FIGURE 4-6 PEV-specific measures for transportation funding. NOTE: PEV, plug-in electric vehicle. SOURCE: Based on data from C2ES (2015). Courtesy of the Center for Climate and Energy Solutions.

untaxed, it could be an effective strategy for addressing the perceived issues around fairness and more clearly elaborating the government's innovation policy by setting criteria like those for other tax incentives found in the U.S. and state tax codes.

Recommendation: Federal and state governments should adopt a PEV innovation policy where PEVs remain free from special roadway or registration surcharges for a limited time to encourage their adoption.

STREAMLINING CODES, PERMITS, AND REGULATIONS

Although there are some applicable federal and state permitting processes that affect PEV infrastructure deployment—such as federal environmental laws (for example, the National Environmental Policy Act, NEPA) and state regulations—cities, counties, and regional governments are at ground-zero for consumer adoption and use of PEVs. Travel distances, trip patterns, and vehicle registration data show that most PEV registrations and travel will be within urbanized areas. The usefulness of the vehicles will largely depend on the availability of charging infrastructure, whether at home, at work, or in public locations (see Chapter 5 for an in-depth discussion of charging infrastructure needs).

Electrical permit requirements appear to vary widely from jurisdiction to jurisdiction (see Table 4-4), as does the amount of time required to apply for and process permits and to obtain a final electrical inspection to certify compliance with applicable electrical codes. Consumer interest in PEVs could be seriously impeded if PEV buyers must bear high permit and installation costs and experience delay in the activation of their home chargers.

Some forward-looking jurisdictions are making adjustments in their electrical codes and permit processes to expedite installation and activation of a home-based charger.[7] Furthermore, many jurisdictions are proactively amending their building codes to require that new construction be "forward compatible" with devices for charging at home (DOE 2014c).

In its interim report, the committee suggested that state and local governments ensure that their permit processes are appropriate for the type of infrastructure project and poten-

[7] Portland, Oregon; Raleigh, North Carolina; and San Francisco, California are three municipalities that have instituted programs to expedite electrical permit processes.

tial impact to the site or broader environment (NRC 2013). There are instances where extensive permit processes and environmental review have been undertaken that would have been appropriate for a highway expansion project but are ill suited for the simple installation of a DC fast-charging station (C2ES 2012). For example, Oregon DOT has reported that even though the DC fast-charging stations installed in Oregon were provided under a master contract by a single vendor, the environmental permit process for each station differed based on the source of funding used to pay the contractor for otherwise identical stations (A. Horvat, Oregon Department of Transportation, personal communication, June 2014). If the charging station was funded with U.S. DOT money through the federal Transportation Investment Generating Economic Recovery (TIGER) grant program, each station was required to undergo heightened NEPA permitting, including an assessment of potential underground hazardous materials. However, if the station was to be funded through DOE, there were no permit requirements beyond those for ordinary state and local permits.

Finding: Regulatory and environmental officials often do not understand the nature, uses, and potential site impacts of charging stations. As a result, unnecessary permit burdens and costs have been introduced to the installation process for public charging stations.

Recommendation: Federal officials should examine current NEPA and other permitting requirements to determine the most appropriate requirements for the class of infrastructure to be installed; the federal government should adopt uniform rules that would apply to all charging installations of a similar asset class, regardless of the capital funding source used to pay for them.

Finding: The permitting and approval processes for home-based and public charging installations need more clarity, predictability, and speed.

Recommendation: Local governments should streamline permitting and adopt building codes that require new construction to be capable of supporting future charging installations. Governments could implement new approaches, perhaps on a trial basis, to learn more about their effectiveness while still ensuring personal and environmental safety.

ANCILLARY INSTITUTIONAL ISSUES RELATED TO SUPPORT FOR PLUG-IN ELECTRIC VEHICLES

Battery Recycling and Disposal

PEV battery recycling and disposal needs will affect the costs and acceptance of PEVs and the infrastructure requirements to support them. At the end of its useful life in the vehicle, the battery must be disposed of, either by applying it to a secondary use (for example, as a back-up power source in a stationary application) or by reusing materials and components that have value and disposing of the remainder as waste. The cost of disposal, less any value in secondary use or of recycled parts and materials, ultimately must be paid by the vehicle owner. Actions that reduce this cost will lower a cost barrier to PEV use.

PEV manufacturers, waste disposal firms, and others are working to create PEV battery recycling and disposal systems. If their efforts lag expansion of the PEV market, it is conceivable that when significant numbers of PEVs begin to reach the end of their lives, a battery-disposal bottleneck could present an obstacle to PEV production and sales. PEV and battery manufacturers have stated that lithium batteries contain no toxic substances that would preclude their disposal in the ordinary waste stream (Kelty 2008; Panasonic 2014). However, because reducing the environmental effects of motor vehicle transportation motivates public support of the PEV market and is attractive to many PEV purchasers, PEV producers have an incentive to develop recycling and reuse options for the batteries.

TABLE 4-4 Variation in Residential Electric Permit Fees by City or State

		Permit Fee ($)		
Region	Number of Permits	Average	Minimum	Maximum
Arizona	66	96.11	26.25	280.80
Los Angeles	109	83.99	45.70	218.76
San Diego	496	213.30	12.00	409.23
San Francisco	401	147.57	29.00	500.00
Tennessee	322	47.15	7.50	108.00
Oregon	316	40.98	12.84	355.04
Washington	497	78.27	27.70	317.25

SOURCE: ECOtality (2013).

In the longer term, recycling of high-value materials or components could be important for restraining PEV battery costs. Although projections indicate that material shortages are unlikely to seriously constrain PEV battery production, large-scale conversion of the fleet to PEVs probably would increase consumption of certain materials, including lithium and cobalt, enough to raise prices significantly. Efficient recycling would moderate material price increases (Gaines and Nelson 2010).

The sections below describe the status of recycling technology; the regulations and standards affecting recycling; prospects for secondary uses of batteries; present involvement of vehicle and battery manufacturers, recycling firms, and others; and possible areas for federal action.

Finding: Reducing the environmental impact of motor vehicle transportation attracts buyer interest and public support for PEVs. Therefore, although the disposal of lithium-ion PEV batteries does not appear to present adverse health risks nor does it have substantial financial advantages, provision for environmentally sound battery disposal will facilitate development of the PEV market.

Recycling Technology

Technologies available today for lithium-ion battery recycling recover certain elementary materials from the battery structure and the cathode, such as cobalt and nickel. The lithium in the cathode is not recovered (ANL 2013; Gaines 2014). Most of the materials obtainable from recycling lithium-ion batteries are of little value compared with the cost of recovery, and newer battery designs that use less expensive materials (in particular, cathodes that do not contain cobalt) yield even less value in recycling. Therefore, recycling is not economical (Kumar 2011; Gaines 2012). Processes under development seek to recover intact, reusable cathode materials that have more value than their elemental components (ANL 2013).

Standards and Regulations

Battery standards are essential for efficient and safe disposal and recycling. Designing batteries with recycling in mind reduces the cost of recycling, and standardization of designs simplifies the operation of recycling facilities. Labeling is necessary to ensure that batteries of different composition can be properly sorted for recycling. Design standards also could facilitate secondary uses.

The Society of Automotive Engineers (SAE) is actively engaged in vehicle electrification standards. Standards under development related to battery disposal include Vehicle Battery Labeling Guidelines (J2936), Identification of Transportation Battery Systems for Recycling Recommended Practice (J2984), Standards for Battery Secondary Use (J3097), and Recommended Practices for Transportation and Handling of Automotive-type Rechargeable Energy Storage Systems (J2950) (SAE International 2014).

No federal or state laws yet require recycling of the batteries contained in PEVs. California and New York require recycling of small rechargeable batteries. In New York, sellers are required to receive used batteries of that type, and battery manufacturers are required to develop plans for collection and recycling. The California law requires sellers to accept used batteries (Gaines 2014). Those laws could provide a pattern for future laws applying to PEV batteries. The federal government regulates the transportation of batteries as hazardous materials (PRBA 2014), but the transport regulations appear to be aimed mainly at the risk of fire from sparks or short circuits.

European Union regulations have established requirements for collection and recycling of all batteries sold to consumers in the European Union. The manufacturer or distributor of the consumer product is responsible for compliance (European Commission 2014).

Finding: Industry standards regarding design and labeling of PEV batteries are necessary for efficient and safe recycling.

Secondary Uses

PEV battery performance (energy storage capacity) declines with use until it becomes unacceptable for powering a vehicle. A battery in this condition, however, might still be usable for other applications, such as energy storage by utilities to satisfy peak demand, storage of energy from an intermittent generator like a solar energy facility, or as backup power in a residence. Developing the market for such secondary uses would reduce the cost of the battery to its initial owner, the PEV purchaser. Reuse delays but does not eliminate the need for eventual recycling or disposal of the battery.

It is most helpful to view battery secondary use (B2U) as an economic ecosystem—a collection of independent stakeholders that could co-evolve around a value chain to bring depleted batteries from the PEV into a secondary system. The maximum potential and limitations of the B2U ecosystem are set by the original design and architecture of the vehicle-battery system. Because the vehicle manufacturers specify the design for the vehicle-battery pack and the parameters for its production, they are currently the most critical player in the development of such an ecosystem. To enable a B2U market to evolve, the vehicle manufacturers must find enough value from participating in the B2U ecosystem to develop a strategy that complements their proprietary PEV technologies.

A B2U strategy must consider the design, development, and manufacture of a battery system with the intent to serve two purposes: (1) the initial use in the vehicle and (2) another application, most likely stationary. An optimal B2U strategy requires the design and use of the battery to maximize the value of the system over its entire extended life cycle. Bowler (2014) developed a model to evaluate trade-offs along the secondary use value chain. The modeling showed that circumstances can exist in which the economic incentives for secondary use become attractive, but this can only be accomplished with the active participation of all the stakeholders in the B2U value chain.

Each vehicle manufacturer could independently develop and use such a model to integrate its own technical parameters into the development of a proprietary B2U strategy. Current evidence suggests that the market will begin with such proprietary deployments. For example, Nissan was first to announce the use of an on-vehicle battery to supplement electric energy to a demonstration home near its headquarters (Pentland 2011). The removal of a depleted PEV battery that had been optimized for stationary use would seem a logical next step. Ford, Tesla, and Toyota have been reported as pursuing various strategies (Woody 2014).

PEV manufacturers are engaged in developing technology and exploring the market for stationary battery applications. Most such efforts are in early stages and include the following examples:

- Nissan Motor Company and Sumitomo Corporation have formed a joint venture (4R Energy Corporation) to store energy from solar generators and other applications using PEV batteries (Srebnik 2012; 4R Energy 2013; Sumitomo 2014). Sumitomo announced installation of a prototype system assembled from 16 used PEV batteries at a solar farm in Japan in February 2014. A battery system has been installed in an apartment building in Tokyo (Nissan Motor Corporation 2013). The venture is working on developing additional applications for used batteries.
- Tesla Motors is supplying batteries to SolarCity, a company that leases and installs solar panels for residential and business customers. The battery is a component of the solar panel system. Trial residential systems were installed in 2013 (Woody 2013). The system is not reported to be reusing PEV batteries but represents a potential market for reuse.
- A Toyota subsidiary (Toyota Turbine) has begun reusing Toyota HEV NiMH batteries in solar panel energy management systems that have been sold to Toyota vehicle dealerships (Toyota Turbine 2013; Nikkei Asian Review 2014).
- General Motors and ABB in 2012 demonstrated a system that packaged five used Chevrolet Volt batteries in a stationary back-up power unit for residential or business applications (General Motors 2012).

Alternatively, the federal government could develop a common public framework that would disseminate information on the actions and processes that create second-use value to the potential participants in a national B2U value chain. That approach might become appropriate as standardization increases among vehicle batteries, charging systems, and the national electric grid.

Finding: Vehicle manufacturers appear to recognize a practical responsibility for disposal of batteries from their vehicles, although their willingness to bear this responsibility voluntarily as PEV sales grow and the fleet ages remains to be seen. Unlike the European Union, the United States imposes no legal requirements for battery disposal on manufacturers or sellers.

Finding: There is a potential market for secondary uses of PEV batteries that are no longer suitable for automotive use but retain a large share of their storage capacity. Whether led by private companies or public agencies, an effective collaboration among the entities that design and manufacture PEVs, the vehicle owners, and the users and purveyors of stationary electric systems can materially assist the development of an economically efficient secondary-use marketplace.

Recycling Arrangements and Capabilities

The principal participants in the PEV battery recycling system will be the vehicle owner, the party that accepts or is required to accept the responsibility for battery disposal (most likely the vehicle manufacturer), companies in the recycling industry, and producers and purchasers of stationary storage units that can reuse PEV batteries. At present, most PEV batteries that have gone out of use probably have passed through PEV dealerships, and manufacturers appear to recognize that they will be expected to provide for battery disposal.

Lead-acid battery recycling is well established in the United States and internationally and is sustained by the value of the recycled lead (that is, recyclers pay for the used batteries they process). Nearly all lead-acid batteries are recycled. The established firms with experience in recycling technology and in the logistics of battery collection, transport, and handling can provide the industrial base for PEV battery recycling (Gaines 2014). The U.S. battery recycling firm Retriev Technologies (until 2013 known as Toxco Industries) recycles lithium-ion PEV batteries (Retriev Technologies 2014). Retriev and the U.K. battery recycling firm Ecobat Technologies are reported to be developing processes for recovery of intact cathode materials from PEV batteries (ANL 2012), a process that has potential for reducing the net cost of battery production and disposal. The Belgian materials and recycling firm Umicore has established a facility in North Carolina to dismantle PEV and HEV batteries before shipment of components to its processing plant in Belgium (Umicore 2014).

Vehicle manufacturers have arrangements with recyclers for battery disposal and have had some involvement in developing improved processes. For example, Tesla has arrangements with recycling companies in Europe and North America for recycling and disposal of used battery packs

(Kelty 2011) and plans to recycle batteries in-house at what it calls the Gigafactory, a battery plant that it intends to build (Tesla 2014).

Finding: The solid waste disposal industry has developed technologies for acceptable disposal of PEV batteries, and technological improvements might succeed in extracting greater net value from recycled materials. However, PEV battery recycling will not pay for itself from the value of recycled materials.

Finding: Battery disposal is not a near-term obstacle to PEV deployment; PEV batteries can be safely disposed of in the general waste stream, and regulating battery disposal at this time could increase the cost of PEV ownership. Thus, federal regulatory action does not appear necessary at this time.

Finding: PEV manufacturers, the solid waste industry, and standards organizations are working to develop disposal, recycling, and reuse technologies. Although federal action is not required, there appear to be opportunities for federal support of industry efforts.

Recommendation: Although battery recycling does not present a barrier to PEVs in the near term, the federal government should monitor the developments in this area and be prepared to engage in research to establish the following: efficient recycling technologies, standards for battery design and labeling that will facilitate safe handling of used batteries and efficient recycling, and regulation to ensure safe transportation and environmentally acceptable disposal of batteries that promotes efficient recycling and avoids creating unintended obstacles.

Emergency Response

Police, firefighters, and emergency medical services (EMS) personnel responding to road crashes that involve PEVs must be aware of the hazards associated with PEVs that differ from the hazards associated with gasoline-powered vehicles in wrecks, and they must be trained in procedures for mitigating these hazards. The hazards are risks of electrical shock, fire, and exposure to toxic substances (NHTSA 2012, p. 2). Because highway emergency response in the United States is the responsibility of thousands of independent local police, fire, and EMS organizations, training and communication of information are challenging activities. All the emergency responders will require training and access to the necessary equipment to discharge batteries safely after an accident and on other safe handling procedures.

The most important nationwide PEV emergency response training activity is Electric Vehicle Safety Training, a project of the National Fire Protection Association (NFPA). NFPA is a nonprofit membership organization engaged in development of codes and standards, training, and research. The training program is funded by a grant from DOE, as part of the department's effort to promote PEV use (NFPA 2014). The NFPA project has developed a variety of training materials and programs and information resources and has conducted a series of courses to train instructors. The NFPA training program is supported by research, involving full-scale testing, to determine best practices for response to incidents involving PEVs. The research has been supported by DOE, the National Highway Traffic Safety Administration (NHTSA), and the automotive industry (Long et al. 2013).

At the federal level, NHTSA develops and distributes EMS training standards and curricula, organizes cooperative activities, maintains databases, and evaluates state EMS systems (NHTSA 2014a). NHTSA has published guidance on safety precautions for vehicle occupants, emergency responders, and towing and repair workers when a PEV is damaged by a collision (NHTSA 2012; NHTSA 2014b). The guidance is brief and general and does not contain detailed technical information or response instructions.

Recommendation: DOE and NHTSA should cooperate in long-term monitoring of the implementation and effectiveness of the NFPA EV Safety Training program. The monitoring should determine whether the program is reaching local emergency responders, whether the skills it teaches prove useful in practice, and whether it is timely.

REFERENCES

ANL (Argonne National Laboratory). 2012. "Advanced Battery Research, Development, and Testing." http://www.transportation.anl.gov/batteries/index.html. Accessed March 14, 2013.

ANL. 2013. "PHEV Batteries." http://www.transportation.anl.gov/batteries/phev_batteries.html.

Battaglia, S. 2013. Are You Willing to Pay an Extra Tax to Drive Your Electric Vehicle? Your Energy Blog, February 28. http://www.yourenergyblog.com/are-you-willing-to-pay-an-extra-tax-to-drive-your-electric-vehicle/. Accessed March 1, 2015.

Bowler, M. 2014. "Battery Second Use: A Framework for Evaluating the Combination of Two Value Chains." Unpublished Doctoral Dissertation, Clemson University.

CAL PEV (California Plug-in Electric Vehicle Collaborative). 2012a. "A Toolkit for Community Plug-in Electric Vehicle Readiness." August. http://www.pevcollaborative.org/sites/all/themes/pev/files/docs/toolkit_final_website.pdf.

CAL PEV. 2012b. "Transportation Funding Consensus Statement." September. http://www.evcollaborative.org/sites/all/themes/pev/files/TransportationFunding_PEVC_ConsensusStatement_120924_Final.pdf.

CBO (Congressional Budget Office). 2012. "How Would Proposed Fuel Economy Standards Affect the Highway

Trust Fund?" May. http://www.cbo.gov/sites/default/files/cbofiles/attachments/05-02-CAFE_brief.pdf.

CBO. 2013. "Testimony on the Status of the Highway Trust Fund before the Subcommittee on Highways and Transit, Committee on Transportation and Infrastructure, and U.S. House of Representatives." Washington, DC. July 23. http://www.cbo.gov/sites/default/files/cbofiles/attachments/44434-HighwayTrustFund_Testimony.pdf.

C2ES (Center for Climate and Energy Solutions). 2012. "Peer Exchange for State DOTs on Electric Vehicles—Charging Infrastructure Workshop." Raleigh, NC, June 25-26. http://www.c2es.org/initiatives/pev/implementing-action-plan/state-dots-electric-vehicles#ChargingInfrastructureWorkshop.

C2ES. 2014. "PEV Action Tool." http://www.c2es.org/pev-action-tool. Accessed October 23, 2014.

C2ES. 2015. "PEV-Specific Measures for Transportation Funding." Accessed March 1, 2015. http://www.c2es.org/initiatives/pev/maps/pev-specific-measures-infrastructure-funding.

D'Artagnan Consulting. 2012. "Washington State Road Usage Charge Assessment: Domestic and International Review and Policy Context." Steering Committee Briefing, September 13. http://waroadusagecharge.files.wordpress.com/2012/08/international-domestic-review_final.pdf.

DOE (U.S. Department of Energy). 2008. "Linkages of DOE's Energy Storage R&D to Batteries and Ultracapacitors for Hybrid, Plug-in Hybrid, and Electric Vehicles." http://www1.eere.energy.gov/ba/pba/pdfs/vehicle_energy_storage_r_and_d_linkages.pdf.

DOE. 2013a. 2013 Annual Merit Review Results Report. http://energy.gov/sites/prod/files/2014/04/f14/2013_amr_02.pdf.

DOE. 2013b. "Energy Innovation Hub: Batteries and Energy Storage." Joint Center for Energy Storage Research. http://science.energy.gov/~/media/bes/pdf/hubs/JCESR_Fact_Sheet.pdf.

DOE. 2013c. "Maps and Data." Alternative Fuels Data Center. http://www.afdc.energy.gov/data/categories/fuel-consumption-and-efficiency#10562.

DOE. 2014a. Fiscal Year 2013 Annual Progress Report for Energy Storage R&D. Publication No. DOE/EE-10-38. Washington, DC. http://energy.gov/eere/vehicles/downloads/vehicle-technologies-office-2013-energy-storage-rd-progress-report-sections.

DOE. 2014b. "Electric Vehicle Community Readiness Projects." http://www1.eere.energy.gov/cleancities/electric_vehicle_projects.html.

DOE. 2014c. "State Laws and Incentives." Alternative Fuels Data Center. http://www.afdc.energy.gov/laws/search?utf8=%E2%9C%93&keyword=building+code&search_by=keyword.

ECOtality. 2013. "The EV Project." http://www.theevproject.com/overview.php. Accessed March 1, 2015.

EPA (U.S. Environmental Protection Agency). 2013. "Light-Duty Automotive Technology and Fuel Economy Trends: 1975 Through 2013." http://www.epa.gov/otaq/fetrends.htm. Accessed March 1, 2015.

European Commission. 2014. "Frequently Asked Questions on Directive 2006/66/EU on Batteries and Accumulators and Waste Batteries and Accumulators." http://ec.europa.eu/environment/waste/batteries/pdf/faq.pdf.

FHWA (Federal Highway Administration). 1997. "Highway Finance." https://www.fhwa.dot.gov/ohim/hs97/hfpage.htm.

FHWA. 2014. "Highway Statistics, 2012. Table HF-10." Office of Highway Policy Information: Highway Statistics Series, March. http://www.fhwa.dot.gov/policyinformation/statistics/2012/hf10.cfm.

4R Energy Corporation. 2013. "Business Model." http://www.4r-energy.com/en/services/. Accessed December 3, 2013.

Gaines, L. 2012. To recycle, or not to recycle, that is the question: Insights from life-cycle analysis. MRS Bulletin 37(4): 333-338.

Gaines, L. 2014. "The Future of Automobile Battery Recycling." Presentation to the Committee on Overcoming Barriers to Electric-Vehicle Deployment, Washington, DC, February 25.

Gaines, L., and P. Nelson. 2010. "Lithium-ion Batteries: Examining Material Demand and Recycling Issues." http://www.transportation.anl.gov/pdfs/B/626.PDF.

GAO (U.S. Government Accountability Office). 2011. Highway Trust Fund: All States Received More Funding Than They Contributed in Highway Taxes from 2005 to 2009. GAO-11-918. Washington, DC, September. http://www.gao.gov/assets/520/511454.pdf.

GAO. 2012a. Batteries and Energy Storage: Federal Initiatives Supported Similar Technologies and Goals but Had Key Differences. GAO-12-842. Washington, DC, August. http://www.gao.gov/assets/650/647742.pdf.

GAO. 2012b. Highway Trust Fund: Pilot Program Could Help Determine Viability of Mileage Fees for Certain Vehicles. GAO 13-77. Washington, DC, December. http://www.gao.gov/assets/660/650863.pdf.

General Motors Corporation. 2012. "GM, ABB Demonstrate Chevrolet Volt Battery Reuse Unit." November. http://media.gm.com/media/us/en/gm/news.detail.html/content/Pages/news/us/en/2012/Nov/electrification/1114_reuse.html.

INL (Idaho National Laboratory). 2014a. "The EV Project." http://avt.inel.gov/evproject.shtml. Accessed October 23, 2014.

INL. 2014b. Observations from The EV Project in Q4 2013. http://avt.inel.gov/pdf/EVProj/EVProjectSummaryReportQ42013FINAL.pdf.

IRS (Internal Revenue Service). 2009. "Notice 2009-89, New Qualified Plug-in Electric Drive Motor Vehicle Credit." Internal Revenue Bulletin 2009-48. http://www.irs.gov/irb/2009-48_IRB/ar09.html.

Kelty, K. 2008. "Mythbusters Part 3: Recycling Our Non-Toxic Battery Packs." Tesla Blog, March 11. http://www.teslamotors.com/blog/mythbusters-part-3-recycling-our-non-toxic-battery-packs.

Kelty, K. 2011. Tesla's Closed-Loop Battery Recycling Program. Tesla Blog, January 26. http://www.teslamotors.com/blog/teslas-closed-loop-battery-recycling-program.

Kumar, A. 2011. The lithium battery recycling challenge. Waste Management World 12(4). http://www.waste-management-world.com/articles/print/volume-12/issue-4/features/the-lithium-battery-recycling-challenge.html.

Long, R., A. Blum, T. Bress, and B. Cotts. 2013. Best Practices for Emergency Response to Incidents Involving Electric Vehicles Battery Hazards: A Report on Full-Scale Testing Results. Fire Protection Research Foundation, June. http://www.nfpa.org/research/fire-protection-research-foundation/reports-and-proceedings/electrical-safety/new-technologies-and-electrical-safety/emergency-response-to-incident-involving-electric-vehicle-battery-hazards.

Michigan Clean Energy Coalition. 2011. "Plug-in Ready Michigan: An Electric Vehicle Preparedness Plan." http://cec-mi.org/wp-content/uploads/2011/11/Plug-In-Ready-Michigan.pdf.

NFPA (National Fire Protection Association). 2014. "About the Project." Electric Vehicle Safety Training. http://www.evsafetytraining.org/about-us.aspx.

NHTSA (National Highway Traffic Safety Administration). 2012. Interim Guidance for Electric and Hybrid-Electric Vehicles Equipped with High Voltage Batteries. DOT HS 811 574. January.

NHTSA. 2014a. "About NHTSA EMS." http://www.ems.gov/mission.htm. Accessed October 23, 2014.

NHTSA. 2014b. "Interim Guidance for Electric and Hybrid-Electric Vehicles Equipped with High Voltage Batteries: Version 2." DOT HS 811 576. Washington, DC. March.

Nikkei Asian Review. 2014. "Toyota Spruces Up Green Credentials with Battery Recycling." January 28. http://asia.nikkei.com/Business/Companies/Toyota-raising-green-profile-by-spearheading-recycling-of-used-HV-batteries.

Nissan Motor Corporation. 2013. "The Compact Lithium-ion Battery." Environmental Activities. http://www.nissan-global.com/EN/ENVIRONMENT/CAR/FUEL_BATTERY/DEVELOPMENT/LITHIUM_ION_BATTERY/. Accessed February 25, 2014.

NRC (National Research Council). 2013. Overcoming Barriers to Electric-Vehicle Deployment: Interim Report. Washington, DC: The National Academies Press.

NSTIFC (National Surface Transportation Infrastructure Financing Commission). 2009. Paying Our Way: A New Framework for Transportation Finance. February. http://financecommission.dot.gov/Documents/NSTIF_Commission_Final_Report_Mar09FNL.pdf.

NSTPRSC (National Surface Transportation Policy and Revenue Study Commission). 2007. Report of the National Surface Transportation Policy and Revenue Study Commission: Transportation for Tomorrow. December. http://transportationfortomorrow.com/final_report/.

ODOT (Oregon Department of Transportation). 2007. "Financial Services Update: Fuels Tax Revenue." ODOT Financial Services, February. http://www.oregon.gov/ODOT/CS/FS/news/vo2-no2_fsupdate.pdf.

Panasonic. 2014. "Panasonic Batteries." Product Information Sheet. http://na.industrial.panasonic.com/products/batteries/rechargeable-batteries/lithium-ion. Accessed March 1, 2015.

Pentland, W. 2011. "Nissan Leaf to Power Homes." Forbes, August 5. http://www.forbes.com/sites/williampentland/2011/08/05/nissan-leaf-to-power-homes/?partner=yahootix.

PRBA (Portable Rechargeable Battery Association). 2014. Information on Batteries in Transport: Packaging and Shipping Batteries Safely. http://www.prba.org/publications/batteries-in-transport-shipping-batteries-safely/.

Rall, J. 2013. Sustainable State Transportation Funding and Spending. National Conference of State Legislatures. NCSL NRI Committee 2014 Spring Webinar Series, April 24. http://www.ncsl.org/documents/transportation/Jamie.pdf.

Retriev Technologies. 2014. "Lithium Ion." http://www.retrievtech.com/recycling/lithium-ion. Accessed March 1, 2015.

Reuters. 2013. "Thomson Reuters Names the World's Top 100 Most Innovative Organizations for 2013." Press Release, October 7. http://thomsonreuters.com/press-releases/102013/thomson-reuters-2013-top-100-global-innovators.

Road Usage Fee Pilot Program. 2013. H.R. 3638, 113th Congress. https://www.congress.gov/bill/113th-congress/house-bill/3638.

Rocky Mountain Institute. 2014. "Project Get Ready." http://www.rmi.org/project_get_ready. Accessed October 23, 2014.

SAE International. 2014. Vehicle Electrification Standards. http://www.sae.org/standardsdev/vehicleelectrification.htm.

Smart, J., and S. White. 2014. "Electric Vehicle Charging Infrastructure Usage Observed in Large-Scale Charging Infrastructure Demonstrations." Presentation to the Committee on Overcoming Barriers to Electric-Vehicle Deployment, Irvine, CA, February 25.

Srebnik, K. 2012. "Nissan Leaf Innovation and Strategy." http://chargedsv.org/wp-content/uploads/2012/08/Srebnik_Charged2012SV_08232012.pdf.

Sumitomo Corporation. 2014. "World's First Large-Scale Power Storage System Made from Reused EV Batteries Completed in Japan." February 7. http://www.sumitomocorp.co.jp/english/news/detail/id=27673.

Tesla. 2014. "Gigafactory." Tesla Blog, February 26. http://www.teslamotors.com/blog/gigafactory.

Toyota Turbine and Systems Inc. 2013. "Enrich Lifestyles by Using Energy Efficiently for the Better Future of the Environment." http://www.toyota-global.com/company/profile/non_automotive_business/new_business_enterprises/pdf/toyota_turbine_and_systems_e.pdf.

TRB (Transportation Research Board). 2006. The Fuel Tax and Alternatives for Transportation Funding: Special Report 285. Washington, DC: The National Academies Press.

TRB. 2011. Equity of Evolving Transportation Finance Mechanisms: Special Report 303. Washington, DC: The National Academies Press.

Umicore. 2014. "Supply: Hybrid Electric Vehicles/Electric Vehicles." http://www.batteryrecycling.umicore.com/supply/HEV_EV/. Accessed October 23, 2014.

Vekshin, A. 2013. "Electric-Car Owners Get Taxed for Not Paying Gas Taxes." Bloomberg BusinessWeek, June 6. http://www.businessweek.com/articles/2013-06-06/electric-car-owners-get-taxed-for-not-paying-gas-taxes.

Weissmann, J. 2012. "Why Your Prius Will Bankrupt Our Highways." Atlantic, February 2. http://www.theatlantic.com/business/archive/2012/02/why-your-prius-will-bankrupt-our-highways/252397/.

Woody, T. 2013. "How Tesla Batteries Are Powering an Energy Revolution." Atlantic, December 5. http://www.theatlantic.com/technology/archive/2013/12/how-tesla-batteries-are-powering-an-energy-revolution/282056/.

Woody, T. 2014. "Car Companies Take Expertise in Battery Power Beyond the Garage." New York Times, March 25.

WSDOT (Washington State Department of Transportation). 2013. "EV Tax and Fee Comparison." Presentation at EV Roadmap Six Conference, Portland, OR, July 30.

WSTC (Washington State Transportation Commission). 2014. Washington State Road Usage Charge Assessment: Business Case Evaluation Final Report. http://waroadusagecharge.files.wordpress.com/2014/01/wa-ruc-business-case-evaluation_01-07-14.pdf.

5

Charging Infrastructure for Plug-in Electric Vehicles

The deployment of plug-in electric vehicles (PEVs) and the fraction of vehicle miles traveled that are fueled by electricity (eVMT) depend critically on charging infrastructure. PEV charging infrastructure (described in Chapter 2) is fundamentally different from the well-developed infrastructure for gasoline fueling. It can be found in a variety of locations, from a PEV owner's home to a workplace to parking lots of restaurants, malls, and airports. A variety of charging options are available, from AC level 1 chargers that use 120 V ac electric circuits that are present in almost every building to DC fast chargers that do not yet have a technology standard. The charging rate also varies from slow (time-insensitive) charging to fast (time-sensitive) charging. Each infrastructure category also has different upfront and ongoing investment costs and returns and different entities that would have an incentive to build such infrastructure, ranging from vehicle owners who might spend about $1,000 to upgrade their home outlet or electric panel to corporations and governments that could spend $100,000 to build a DC fast-charging station. The public charging stations might also require technology to monitor usage and bill customers. PEV deployment and eVMT will be constrained if charging infrastructure is not conveniently located or if the available infrastructure does not facilitate charging within a convenient time frame. Thus, critical questions for vehicle manufacturers and policy makers are how are vehicle deployment and eVMT affected by the availability of various charging infrastructure types and what is the cost effectiveness of infrastructure investments relative to other investments that manufacturers and the government could make to overcome barriers to PEV deployment.

This chapter considers scenarios for deploying PEV charging infrastructure and the potential effect of that infrastructure on PEV deployment and eVMT. The committee has categorized infrastructure by location (home, workplace, intracity, intercity, and interstate) and power (AC level 1, AC level 2, and DC fast). The infrastructure categories are ranked in order of importance for increasing PEV deployment and eVMT from the perspective of owners of the four PEV classes as defined in Chapter 2. The experience and needs of current early adopters were considered by the committee, but deployment scenarios are focused on mainstream PEV deployment. The chapter concludes by considering which entities might have an incentive to build each category of charging infrastructure, with particular attention to how infrastructure investments would be recovered. The committee provides its findings and recommendations throughout this chapter.

In this chapter, the committee's analysis of infrastructure deployment assumes (1) no disruptive changes to current PEV performance and only gradual improvements in battery capacity over time, (2) early majority buyers who do not plan to make changes to their lifestyles to acquire a PEV, (3) electricity costs that are significantly less expensive than those of gasoline per mile of travel, and (4) a cost for public and workplace charging that is at least as high as that for home charging. The committee notes that the need for charging infrastructure could conceivably be mitigated by investments in battery swapping stations, which use robotic processes and allow drivers to swap batteries in less than 3 minutes. The first major initiative for battery swapping services was launched by Better Place, which built networks of stations in Israel and Denmark but declared bankruptcy in May 2013. Tesla has announced a plan to add battery-swap technology at its network of fast-charging stations (Vance 2013). However, this model is not widely available at this time and is not discussed further in this report.

CHARGING INFRASTRUCTURE AND EFFECTS ON DEPLOYMENT OF PLUG-IN ELECTRIC VEHICLES AND ON ELECTRIC VEHICLE MILES TRAVELED

As discussed in Chapter 2, today's charging infrastructure technology consists of AC level 1 and AC level 2 chargers, which are typically used when charging time is not a prime consideration, and DC fast chargers, which are typically used when charging time is an important consideration. All PEVs can charge with AC level 1 and level 2 chargers, and most battery electric vehicles (BEVs) can also charge at DC fast chargers. In the future, some plug-in hybrid electric vehicles (PHEVs) might be equipped to use DC fast chargers, but there is little motivation to make such a change because PHEVs can use their internal-combustion engines (ICEs) to circumvent the need to charge. Charging infrastructure locations and investments range widely from an existing extensive network of private chargers (or simply ordinary outlets) at homes and workplaces to an expanding infrastructure of public chargers, such as those at retailers or shopping malls or along highways. Workplace and public charging infrastructure might require payment for electricity or time occupying the charger or be restricted to vehicles belonging to a subscription plan or to a certain vehicle manufacturer.

In the mature market, the ideal number, location, and type of charging infrastructure will depend on the demand for different types of PEVs, their use, and their geographic distribution. Conversely, although there has been little research on the relationship between charging-station deployment and PEV deployment, the availability of charging infrastructure and the rate of its deployment might itself influence PEV deployment and use. Figure 5-1 shows six categories of charging-infrastructure deployment, ranked in a pyramid that reflects their relative importance as assessed by the committee. As noted above, the categories are defined by location and power. The term *intercity* refers to travel over distances less than twice the range of limited-range BEVs, and *interstate* refers to travel over longer distances.

Table 5-1 provides the committee's assessment of the effect of charging infrastructure on different PEV classes. Evaluating infrastructure by type of PEV might help to address misconceptions about charging infrastructure needs. For example, PHEVs do not require electric charging for range extension because drivers have the option of fueling with gasoline. BEVs, which have only electricity as a fuel option, are much more affected by the availability of charging infrastructure. That does not mean that electric-charging infrastructure is not important for PHEV deployment, however. PHEV drivers might still heavily use charging at private and public locations to maximize their value proposition in terms of cheaper charging, convenience, or personal values, such as environmental concerns. For example, data from the EV Project on early adopters of the Chevrolet Volt show that 14 percent of charging events occurred away from home, which is similar to the percentage of charging away from home (16 percent) for Nissan Leaf drivers (ECOtality 2013; Smart and Schey 2012). Each charging-infrastructure category and the impact of each category on different PEV classes are discussed in detail in the sections below.

Home Charging

Home charging is a virtual necessity for mainstream PEV buyers of all four vehicle classes given that the vehicle is typically parked at a residence for the longest portion of the day. As shown in Figure 5-2, the U.S. vehicle fleet spends about 80 percent of its time parked at home, and more than 50 percent of the U.S. vehicle fleet is parked at home even during weekday work hours. Most early adopters of PEVs have satisfied their charging needs primarily by plugging their vehicles into 120 V (AC level 1) or 240 V (AC level 2) receptacles at home during overnight hours or other periods when it is convenient to leave their vehicles idle. Even the

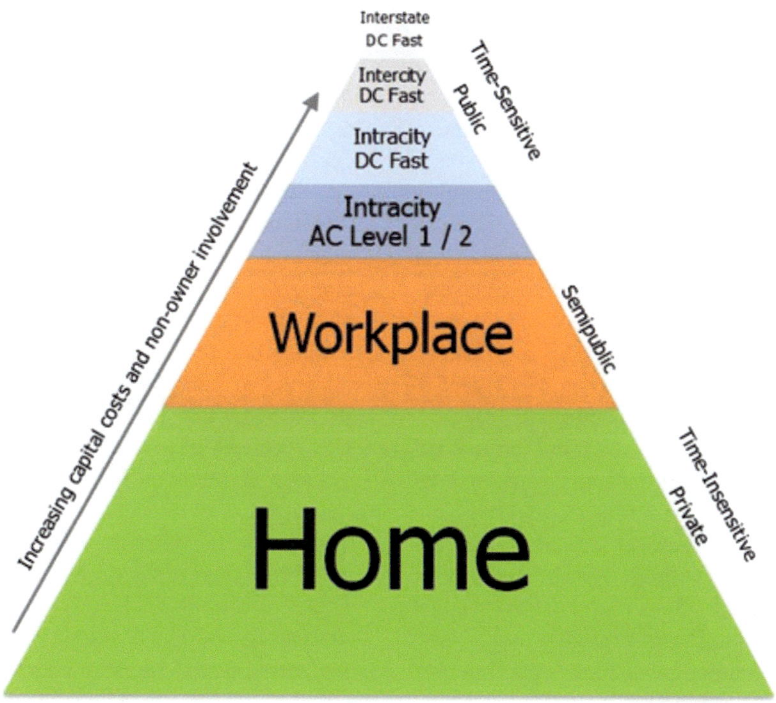

FIGURE 5-1 PEV charging infrastructure categories, ranked by their likely importance to PEV deployment, with the most important, home charging, on the bottom, and the least important, interstate DC fast charging, at the top. NOTE: AC, alternating current; DC, direct current.

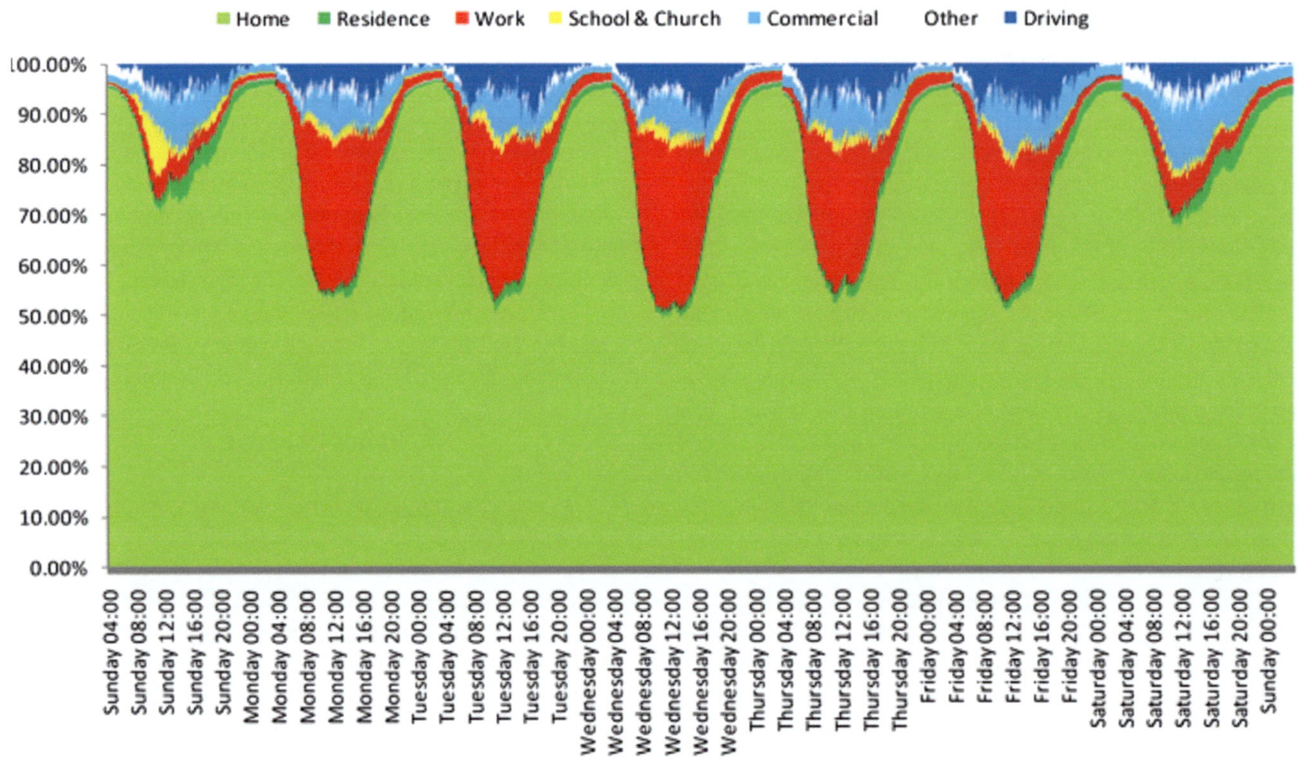

FIGURE 5-2 Vehicle locations throughout the week on the basis of data from the 2001 National Household Travel Survey. SOURCE: Tate and Savagian (2009). Reprinted with permission from SAE paper 2009-01-1311 Copyright © 2009 SAE International.

large 85 kWh battery in a Model S can be fully charged overnight with the 10 kW AC level 2 charger recommended by Tesla for home use. A full battery charge will not usually be needed each night because such charging will typically replace only the electricity used for the previous day's driving. For typical daily trip distances, only a few hours of charging will be required for all types of PEVs.

Home charging is a paradigm shift in refueling behavior for drivers accustomed to refueling quickly at gasoline stations. Many find home charging more convenient than refueling at public stations. For example, in the EV Project study, about 85 percent of Volt charging events and 80 percent of Leaf charging events occurred at home (Smart 2014a).

Home-charging infrastructure is not a barrier to PEV deployment for households with a dedicated parking spot with an electric outlet nearby. According to the U.S. Census Bureau (2011a), nearly two-thirds of U.S. housing structures have garages or carports.[1] Similarly, a representative telephone survey of 1,004 U.S. adults found that 84 percent of respondents had dedicated off-street parking and 52 percent of respondents had a garage or dedicated parking spot with access to an outlet (Consumers Union and the Union of Concerned Scientists 2013). Traut et al. (2013) used data from the U.S. Census and the U.S. Department of Energy (DOE) Residential Energy Consumption Survey to estimate the potential for residential charging of PEVs using various assumptions about missing data on, for example, the presence and size of driveways, the usability of electric outlets, and the number of parking spaces actually available for parking. Although 79 percent of U.S. households have dedicated off-street parking, many households have multiple vehicles, and under base-case assumptions, only 56 percent of vehicles have dedicated off-street parking, and only 47 percent at an owned residence. Additionally, although 38 percent of all U.S. households are estimated to have charging access for at least some vehicles, only an estimated 22 percent of all U.S. vehicles have a dedicated home parking space within reach of an outlet sufficient to recharge a small PEV battery overnight.

Given the number of households with access to dedicated parking with an outlet, PEVs could become a much larger share of the U.S. vehicle market while still relying on ubiquitous residential circuits to accommodate most charging needs. Given the large number of households that do not yet drive PEVs and could take advantage of the convenience of charging at home, the scenario that seems most likely to emerge over the next decade is one in which the growth of demand for PEVs comes primarily from households who

[1] Some of the structures accommodate multiple households.

TABLE 5-1 Effect of Charging-Infrastructure Categories on Mainstream PEV Owners by PEV Class[a]

Infrastructure Category	PEV Class	Effect of Infrastructure on Mainstream PEV Owners
Interstate DC fast charge	Long-range BEV	Range extension, expands market
	Limited-range BEV	Not practical for long trips
	Range-extended PHEV	NA – not equipped
	Minimal PHEV	NA – not equipped
Intercity DC fast charge[b]	Long-range BEV	Range extension, expands market
	Limited-range BEV	2 × Range extension, increases confidence
	Range-extended PHEV	NA – not equipped
	Minimal PHEV	NA – not equipped
Intracity DC fast charge[b]	Long-range BEV	Not necessary
	Limited-range BEV	Range extension, increases confidence
	Range-extended PHEV	NA – not equipped
	Minimal PHEV	NA – not equipped
Intracity AC levels 1 and 2[b]	Long-range BEV	Not necessary
	Limited-range BEV	Range extension, increases confidence
	Range-extended PHEV	Increases eVMT and value proposition
	Minimal PHEV	Increases eVMT and value proposition
Workplace	Long-range BEV	Range extension, expands market
	Limited-range BEV	Range extension, expands market
	Range-extended PHEV	Increases eVMT and value proposition; expands market
	Minimal PHEV	Increases eVMT and value proposition; expands market
Home	Long-range BEV	Virtual necessity
	Limited-range BEV	Virtual necessity
	Range-extended PHEV	Virtual necessity
	Minimal PHEV	Virtual necessity

[a] Assumptions in this analysis are that electricity costs would be cheaper than gasoline costs, that away-from-home charging would generally cost as much as or more than home charging, that people would not plan to change their mobility needs to acquire a PEV, and that there would be no disruptive changes to current PEV performance and only gradual improvements in battery capacity over time.

[b] It is possible that these infrastructure categories could expand the market for the various types of PEVs as appropriate, but that link is more tenuous than the cases noted in the table for other infrastructure categories.

NOTE: AC, alternating current; BEV, battery electric vehicle; DC, direct current; eVMT, electric vehicle miles traveled; NA, not applicable; PEV, plug-in electric vehicle; PHEV, plug-in hybrid electric vehicle.

intend to meet their charging needs predominantly through slow charging at home.

Lack of access to charging infrastructure at home will constitute a significant barrier to PEV deployment for households without a dedicated parking spot or for whom the parking location is far from access to electricity. Those demographic groups include many owners and renters of housing in multifamily dwellings and many households in large cities with on-street parking. About 25 percent of U.S. households live in multifamily residential complexes (U.S. Census Bureau 2011b), and the telephone survey noted above indicated that although 61 percent of single-family houses had access to charging, only 27 percent of multifamily dwellings had parking spaces with access to charging (Consumers Union and the Union of Concerned Scientists 2013). Multifamily residential complexes can face many challenges in installing PEV charging equipment; some are similar to a typical commercial building, and others are unique to multifamily dwellings. Similar to commercial buildings, the electrical panel might be far from the desired charging location, and installation can therefore be costly.

Unique to multifamily residential complexes are the ownership, responsibility, liability, and control of each individual parking space. Multifamily residential complexes have many ways to assign parking to their residents, including dedicated, shared, and leased parking. For residents who have dedicated spaces, the main challenges besides the installation costs are questions within the governance structure for multifamily residential complexes concerning (1) who should bear the cost of upgrading the main panel (if needed) and (2) who will pay for the electricity for charging the PEV. Those costs can be prohibitive for an individual consumer if he or she is responsible for upgrading service to the main panel for the multifamily dwelling. For residents who have shared spaces, additional questions need to be resolved within the governance structure of the multifamily residential complex concerning installation costs, use of charge-enabled spaces, and payment for the electricity. Because no charging space is dedicated to a specific resident, an individual is discouraged from investing in the installation of a charging station because that would not necessarily guarantee him or her the right to use it. In addition, the use of the charging station can no longer be tied to an individual and raises the question of who should pay for the electricity. Lastly, for leased or rented spaces, there is the question of ownership of the PEV charging equipment: which entity should pay for the PEV charging equipment and how should liability be assigned? If tenants are liable for all upgrades, they have a disincentive to perform the upgrades because they might leave. If the owners are liable for all upgrades, they have a disincentive to install them unless they can charge a premium for them or otherwise be compensated.

For residents who do not have any parking available and must rely on on-street parking, the same challenges exist except that the owner or deciding body is not the multifamily residential complex. Instead, it is the local city government that must make policy decisions surrounding installation and operation of PEV charging equipment (Peterson 2011).

Lack of home charging at multifamily complexes or in neighborhoods with on-street parking is a barrier to deployment for owners of all types of PEVs, but most importantly BEVs, particularly limited-range BEVs for which daily charging cannot, like PHEVs, be replaced with gasoline or, like long-range BEVs, postponed. It is also a barrier to increased eVMT for all PEV owners. Overcoming lack of home charging at multifamily residential complexes and in neighborhoods with on-street parking requires providing such consumers with designated parking spaces to charge their vehicles during prolonged times when their vehicles are not in use, such as at workplaces. Although retrofits of multifamily housing for PEV charging might be difficult, facilitating installation of home-charging infrastructure can be accomplished by preparing the sites for installation during initial construction. California mandatory building codes will require new multifamily dwellings to be capable of supporting future charging installations (DOE 2014a).[2] Additionally, multifamily dwelling owners might choose to contract with a charging provider to facilitate installation and payment for charging services. Another interesting model for extending PEV driving to households without access to home charging is to deploy PEVs in car-sharing fleets. That approach is particularly important for the large portion of multifamily dwelling residents who are not in the new vehicle market as compared with single-family home residents. Car sharing is discussed from a consumer perspective in Chapter 3.

Finding: Homes are and will likely remain the most important location for charging infrastructure.

Finding: Lack of access to charging infrastructure for residents of multifamily dwellings is a barrier that will need to be overcome to promote PEV deployment to that segment.

Workplace Charging

Charging at workplaces provides an important opportunity to encourage the adoption of PEVs and increase eVMT. BEV drivers could potentially double their daily range as long as their vehicles could be fully charged both at work and at home, and PHEV drivers could potentially double their all-electric miles. Extending the electric range of PHEVs with workplace charging improves the value proposition for PHEV drivers because electric fueling is less expensive than gasoline. For BEVs and PHEVs, workplace charging could expand the number of people whose needs could be served by a PEV, thereby expanding the market for PEVs. Workplace charging might also allow households that lack access to residential charging the opportunity to commute with a PEV. Furthermore, Peterson and Michalek (2013) estimated that installing workplace charging was more cost-effective than installing public charging; however, it should be noted that installing workplace or public charging was substantially less cost effective than improving the all-electric range of a vehicle.

Data from early adopters in the EV Project shows that workplace charging is used when it is available (Table 5-2). Specifically, Nissan Leaf drivers who had access to workplace charging obtained 30 percent of their charging energy at work, and Chevrolet Volt drivers who had access to workplace charging obtained 37 percent at work. Furthermore, there is some evidence that workplace charging enables longer routine commutes or more daily miles. Of Nissan Leafs that had workplace charging, 14 percent routinely required workplace charging to complete their daily mileage (at least 50 percent of days), but another 43 percent of the Leaf vehicles required workplace charging to complete their daily miles on some days (at least 5 percent of days). Moreover, Nissan Leaf drivers extended their range by 15 miles or 26 percent on days when charging was needed to complete their trips (such days averaged 73 miles traveled) and by 12 percent on days when they charged even though a charge was not required to complete their trips (Smart 2014b).

In considering whether to provide workplace charging, employers confront a number of challenges. One set of chal-

[2] For an explanation of these codes, see California Green Building Code A4.106.8.2 and California AB 1092.

TABLE 5-2 Charging Patterns for Nissan Leafs and Chevrolet Volts

Vehicle	Percent Charging Energy Obtained at Various Locations		
	Home	Work	Other
All Drivers			
Nissan Leaf	86	—	14
Chevrolet Volt	85	—	15
Drivers with Access to Workplace Charging			
Nissan Leaf (~12%)[a]	68	30	2
Chevrolet Volt (~5%)[a]	60	37	3

[a] Numbers in parentheses are percentage of drivers known to have access to workplace charging.
SOURCE: Based on data from ECOtality (2014a,b).

lenges is to determine the rate of PEV adoption by employees, what level of charging would be sufficient for their needs, and how access to chargers can be ensured as the number of PEVs increases. A worker who relies on workplace charging of a BEV might not be able to return home if no charger is available. There is also the possibility that electricity provided to employees will have to be paid for by the employees or taxed as income (IRS 2014).[3] A requirement to assess the value of the charging or report the imputed income could be an impediment to workplace charging. Yet another potential impediment arises from the surcharges that utilities impose on companies that draw more than a threshold level of power. Such demand charges (discussed in Chapter 6) can be substantial.

Workplace charging is becoming available at a small but growing number of companies that offer it as a way of attracting and retaining employees and as a way of distinguishing themselves as green companies.[4] It is an attractive perk if the employer provides charging for the same price or less than is available at home. In assessing the reasons for offering workplace charging, some employers anticipate that concerns about carbon emissions from commuting will eventually generate much stronger pressures for workplace charging and are attempting to move expeditiously by expanding their network of charging stations now (Ahmed 2013). Because of the costs involved and the fact that adding a charging station leaves fewer parking spaces available for employees who do not drive PEVs, Cisco has a policy of increasing the number of workplace charging stations in proportion to the number of employees who express an interest in using them. This tends to have positive feedback effects as increases in the number employees who use workplace charging stations stimulate other employees' interests in acquiring PEVs (Jennings 2013), thereby contributing to a continuing expansion in the number of workplace chargers. Other firms, however, have been reluctant to provide workplace charging on grounds of equity, expressing concerns about providing a perk that would benefit only a relatively small number of employees, at least initially (Musgrove 2013). Recognizing workplace charging as an important opportunity to expand PEV deployment and eVMT, DOE supports the EV Everywhere Workplace Charging Challenge. The Workplace Charging Challenge and the Clean Cities program both provide several guides and resources for employers to simplify the process of adding workplace charging (DOE 2014b; DOE 2013).

Finding: Workplace charging could be an alternative to home charging for those who do not have access to charging infrastructure at home.

Finding: Charging at workplaces provides an important opportunity to encourage PEV adoption and increase the fraction of miles that are fueled by electricity.

Finding: The administrative cost to assess the value of charging or report the imputed income could be an impediment to workplaces to install charging.

Recommendation: The federal government should explicitly address whether the provision of workplace charging at the expense of employers should be included in the recipient's pay or regarded as a benefit that is exempted from taxation.

Public Charging Infrastructure

A critical question to answer is whether lack of public

[3] IRS Publication 15-B states that any fringe benefit is taxable and must be included in the recipient's pay unless the law explicitly excludes it. Although exclusions currently apply to many fringe benefits, the issue of excluding electricity that employers provide at workplace chargers has apparently not yet been explicitly addressed. The issue does not arise at workplaces that engage an outside entity (the installer of the charging infrastructure) to manage the charging units and collect a monthly fee from workers who use them.

[4] To facilitate the process, the Department of Energy (DOE), under the Workplace Charging Challenge launched in January 2013, offers various resources to interested employers, building owners, employees, and others. The resources include information about PEVs, their charging needs, and activities that DOE and communities across the country are doing to support PEV deployment.

charging infrastructure is a barrier to PEV deployment.[5] As shown in Figure 5-1, home charging infrastructure is and is expected to remain more convenient and more critical to PEV deployment than public charging infrastructure. There is no consensus in the research and policy communities, however, on the impact of public charging infrastructure on PEV deployment. Experience in Japan indicates that increased availability of public charging stations reduces range anxiety and leads to more miles driven by BEVs. For example, the building of a single additional fast charger for a TEPCO fleet of BEVs increased eVMT from 203 km/month to 1,472 km/month. Interestingly, no additional energy consumption from the public charger was observed after building the second charger, but drivers allowed their state of charge to go below 50 percent, a sign that their fear of running out of charge had been alleviated (Anegawa 2010).

DOE (2015) estimates that there were more than 9,300 public charging stations in the United States as of April 2015; many stations, however, are only accessible to members of associated subscription-based plans or to vehicles produced by individual manufacturers. Interactive maps of charging stations are updated frequently on the DOE Alternative Fuels Database and through the PlugShare website (DOE 2015; Recargo 2014). Nearly 8,700 of the public charging stations provide AC level 2 chargers, which can add about 10-20 miles of range to a vehicle for each hour of charging, depending on the model and driving conditions. More than 800 public DC fast-charging stations had also been installed by April 2015 (DOE 2015). Networks of DC fast chargers have been installed in Washington, Oregon, and California; along the East Coast I-95 corridor; and the "Tennessee Triangle," which connects Nashville, Chattanooga, and Knoxville. Clusters of DC fast chargers are also in Dallas-Fort Worth, Houston, Phoenix, Atlanta, Chicago, and Southern Florida. Tesla and Nissan Motors—manufacturers of the vehicles that have led BEV sales in the United States—have been actively engaged in expanding their networks of fast chargers. In fact, most of the chargers outside of the regions noted above are part of the proprietary Tesla network of Superchargers (see Chapter 2, Figure 2-10). Tesla had installed more than 190 charging stations in the continental United States and Canada by April 2015 and has plans to expand its network to several hundred stations by the end of 2015, with the stated goal that 98 percent of U.S. drivers are within 100 miles of a Supercharger by 2015 (Tesla 2014). Nissan has announced plans to add at least 500 fast-charging stations by mid-2015 and has partnered with CarCharging to expand networks in California and on the East Coast and with NRG/eVgo to develop a network in the Washington, D.C. area (CarCharging 2013; Nissan 2013).

Several studies have modeled optimal numbers and locations of PEV charging sites from the perspective of limited-range BEV drivers, who have the greatest need for charging. One study looked at the locations where light-duty vehicles parked and modeled optimal charging locations assuming similar trip needs for PEV drivers and ICE drivers (Chen et al. 2013). Other studies have examined trip diary data from such cities as Seattle and Chicago and such states as California to see which trips were not likely to be completed with today's BEVs and sought to place chargers to allow completion of these "failed" trips. Models were optimized by minimizing time or distance deviations from trips required to drive to charging locations. The study of California drivers found that with an 80-mile limited-range BEV, 71.2 percent of the total miles driven and 95 percent of trips could be completed with no public charging required. Optimal placement of 200 DC fast chargers in the state would allow those drivers to complete over 90 percent of miles with two or fewer charges (Nicholas et al. 2013). The data from Chicago and Seattle metro areas showed that no public charging was needed to complete 94 percent and 97 percent of trips, respectively, and optimally locating 100 or 50 stations with 10 AC level 2 chargers each in Chicago or Seattle resulted in mean route deviations of only 1.6 and 0.3 miles, respectively, to make the remaining trips (Andrews et al. 2013). As noted, most studies have not investigated the effect of charging infrastructure deployment on vehicle deployment.

The majority of public charging stations are not yet heavily used. For example, public DC fast chargers in the EV Project were occupied on average 2.3 percent of the time from October-December 2013, and public AC level 2 chargers were occupied 5.5 percent of the time on average (INL 2014). Despite that low utilization, it is not unusual at some popular stations for drivers to have to wait for a charging plug to become available. In addressing the adequacy of the existing network of public charging infrastructure, it is important to understand the factors that contribute to both overutilization and underutilization. The factors include the ratio of charging stations to PEVs in any given area, the location of charging stations, the cost of using the stations, the amount of time it takes to recharge, and restrictions on station use associated with either subscription-membership requirements or incompatible hardware. Low utilization of the charging stations in a given area does not necessarily imply that the network of charging infrastructure is adequate and could instead reflect any combination of the factors noted. Similarly, queuing at charging stations does not necessarily imply that more charging stations should be built, but it is unlikely that most potential customers would be willing to wait for multiple charges to be completed. To the extent that the demand to use charging stations is not uniformly distributed over time and that investments in charging stations are costly, a certain degree of queuing is inherent in a network of charging stations that optimally balances the cost of waiting to charge against the cost of building more charging stations. In addition, at stations that do not impose usage fees or charges for electricity consumed, queuing might partly reflect the fact

[5] The term *public charging infrastructure* refers to charging infrastructure that is located in public spaces but does not imply that the services are offered for free.

that using those stations is cheaper than charging at home. For some locations, such as retail establishments, medical facilities, and commercial parking lots, for-pay AC level 2 infrastructure is used more frequently than free public AC level 2 infrastructure; this might indicate better siting of or more chargers to reduce queueing at for-pay infrastructure (Smart and White 2014).

Over the course of its study, the committee heard concerns that public funding combined with pressures to install public infrastructure quickly has led to some poor siting decisions. So, the fundamental questions remain—how much public infrastructure is needed and where should it be located? There are many complexities associated with installing public charging infrastructure that need to be considered. It can be located within cities, such as at malls or parking lots, or along interstate highways or other corridors. It can include AC level 1, AC level 2, and DC fast charging. It can be costly to install and maintain, and its effect on deployment and eVMT remains unclear, although it enables PEV drivers to extend the electric range of their vehicles beyond the mileage that can be driven on a single charge and might encourage the adoption of limited-range BEVs by mitigating concerns about becoming stranded. However, a substantial amount of public charging infrastructure that is obviously unused could become a symbol that PEVs are not as practical as had been hoped. The following sections consider the location of public infrastructure and its effects on PEV deployment and eVMT.

Finding: Public charging infrastructure has the potential to provide range confidence and extend the range for limited-range BEV drivers, to allow long-distance travel for long-range BEV drivers, and to increase eVMT and the value proposition for PHEV drivers.

Finding: More research and market experience are needed to determine how much public infrastructure is needed and where it should be sited to promote PEV deployment and to encourage PEV owners to optimize vehicle usage.

Recommendation: The federal government through the Departments of Energy or Transportation should sponsor research to study the impact of the public charging infrastructure, including the extent to which its availability affects PEV adoption.

Intracity AC Level 1 and Level 2 Charging Infrastructure

Public AC level 1 and level 2 chargers are now available in some cities, especially where PEV deployment has been relatively strong. Because AC level 1 chargers provide about 4-5 miles of operation per hour of charge, they could be used when charging time is not a primary concern, such as at airports and train stations, where people park their cars for prolonged periods. They can also be installed easily using accessible 120 V outlets. AC level 2 chargers are also becoming increasingly available at locations where vehicles are often parked for just an hour or two, such as at shopping malls, museums, libraries, and restaurants. Installation of chargers at those locations is often seen as a way for businesses to attract customers. Charging providers are also installing AC level 1 and AC level 2 within cities as part of their subscription-based business model. Some utilities are also installing infrastructure and are motivated to provide public charging to encourage PEV deployment and hence sell more electricity to residential customers with PEVs. Infrastructure-deployment models are discussed in more detail at the conclusion of this chapter.

Although the committee did not attempt to establish guidelines for locating public charging infrastructure, it seems reasonable to assume that to maximize the use of intracity charging infrastructure, chargers must be dispersed around metropolitan areas and placed at convenient locations. Siting of public charging stations is driven by a variety of motivations, and the stations are operated by both public and for-profit entities. Charging providers might locate public stations to maximize revenue from for-pay stations, to establish their image as a green business or government, to induce customers to stop at their establishments, to take advantage of favorable conditions (such as no-cost land or easy access to electricity source), to increase deployment of vehicles, to increase eVMT, or to relieve range anxiety. Data from intracity AC level 2 infrastructure associated with the EV Project indicate that chargers located at parking lots and garages, transportation hubs, workplaces, and public or municipal sites were used most frequently. Least frequently used sites were at educational institutions, multifamily residences, and medical facilities (Smart 2014c).

The effects of intracity AC level 1 and level 2 charging infrastructure vary by PEV class as seen in Table 5-1. Long-range BEVs will have little use for slow charging in public locations as there will be little value of charging slowly given their sufficient all-electric range. However, they might choose to top-off their charge when convenient or if perks, such as free parking at an airport, are available. Limited-range BEVs are expected to experience the most utility from intracity AC level 1 and level 2 charging by assuring them that they will not be stranded if their charge is depleted and by allowing them to extend their daily mileage beyond a full battery charge. With limited battery ranges and no other choice for fuel, charging in public is an attractive option for limited-range BEVs. Both minimal and extended-range PHEVs are predicted to use intracity AC level 1 and level 2 charging for increased eVMT and hence to realize an increased value proposition of their vehicles. However, they do not need intracity chargers for range extension or range confidence because they can also fuel on gasoline. Increased eVMT from charging in public might be particularly useful for minimal PHEVs whose smaller batteries could be nearly fully charged in a shorter time, thus extending their small ranges substantially if they are able to charge frequently throughout the day.

Intracity DC Fast-Charging Infrastructure

DC fast-charging technology was described in Chapter 2. Although DC fast chargers are often considered for corridor travel, such as between cities or states, the majority of the fast-charge infrastructure is installed within cities and their metro areas. There are some data to indicate that BEV owners prefer fast charging to complete a journey or otherwise to create options for using the vehicle beyond its routine range. EV Project data on the percent of DC fast charges that occurred on trips of a given length provide information on charging behavior of Nissan Leaf drivers (Smart and White 2014; J. Smart, Idaho National Laboratory, personal communication, November 6, 2014). In the fourth quarter of 2013, after the institution of fees to charge at some DC fast-charging locations, 56 percent of outings that included a fast charge were greater than 60 miles round trip, and 44 percent of outings that included a fast charge were less than 60 miles round trip. Some of the less than 60 mile round-trips that included a DC fast charge might reflect the value a driver places on a DC fast charge even when it is not required to complete the trip. However, many of the short trips (63 percent) started with a less than full battery, indicating that the charge might have been required to return home. When an outing included a DC fast charge and began with a full battery, average round trip distance was 87.5 miles. That observation again indicates that many trips that include a DC fast charge required a charge to complete, and DC fast charging might have been the most convenient way to acquire the charge.

The impact of intracity DC fast-charging infrastructure varies by PEV type, as noted in Table 5-1. Long-range BEVs will have little use for fast charging in cities as their vehicle range is unlikely to require range extension or range confidence. However, charging at a DC fast-charging station would allow them to acquire a full battery charge more quickly than home charging; this option might be valuable to a long-range BEV owner, particularly one who does not have a place to charge at home. The committee notes that Tesla—the only current producer of a long-range BEV—is implementing a model in which charging at its DC fast charger stations is included in the price of the vehicle. Limited-range BEVs are expected to experience the most utility from intracity DC fast charging as it provides range confidence that they will not be stranded and range extension in less time than that required for AC level 1 or level 2 charging. In April 2014, Nissan began offering new Leaf buyers in several markets free public charging through a special card that allows using several charging providers. Range-extended and minimal PHEVs are unable to use DC fast-charging infrastructure, so this segment of infrastructure deployment does not apply to PHEV owners.

Intercity and Interstate DC Fast-Charging Infrastructure

The availability of DC fast chargers along highways connecting cities and states has facilitated regional travel for limited-range BEVs and enabled long-distance travel for long-range BEVs. An example of such a network is the corridor of DC fast chargers installed at about 40-mile intervals along Interstate 5 in Washington and Oregon. Such infrastructure provides long-range BEVs with multiple places to acquire a charge on an extended trip and enables limited-range BEVs to travel between two cities in the same region. For travel between cities where stops to charge might be inconvenient, DC fast chargers are expected to be used primarily for range extension and are expected to receive less use than DC fast chargers within cities. Although data from the EV Project is primarily from cities, a preliminary study of charging along the I-5 corridor shows that most charges do in fact occur within cities rather than between them (Smart 2014d). Although some early adopters of limited-range BEVs have chosen to drive their vehicles long distances requiring multiple battery charges, the committee's view is that the vast majority of limited-range BEV drivers will restrict themselves to a range that requires at most one full charge between neighboring cities. As noted, PHEVs are not equipped to use DC fast-charging stations and can extend their range by refueling on gasoline.

Thus, interstate DC fast chargers are projected to be the least important type of infrastructure for PEVs because it will not (or cannot) be used by PHEVs and will be inconvenient for limited-range BEVs. However, it should be noted that there are alternative scenarios in which interstate DC fast chargers do become an important type of infrastructure. An example of such a scenario is if the market becomes dominated by long-range BEVs that are used as primary vehicles. If that is the case, home charging infrastructure will continue to be most important for drivers' everyday usage, and workplace and intracity infrastructure will be relatively unimportant. Intercity and interstate charging would, in that scenario, enable long-range BEVs to take longer trips with relative ease. Vehicle manufacturers, especially those focused on BEVs, are building intracity, intercity, and interstate DC fast-charging infrastructure; this indicates that they think it is valuable. It is not clear whether they are doing this for marketing or business strategy reasons or to spur vehicle deployment in the near term or whether they believe that this type of infrastructure will be necessary in the future.

MODELS FOR INFRASTRUCTURE DEPLOYMENT

To understand how best to overcome any infrastructure barriers to PEV deployment, one must consider the installation and operating costs for the different categories of charging infrastructure, the possible deployment models, and who might have an incentive to build such infrastructure. Several different entities might have an incentive to build or operate charging infrastructure; these include vehicle owners, workplaces, retailers, charging providers, utilities, vehicle manufacturers, and the government. Their motivations might include generating revenue, improving air quality, selling more electricity, or selling more PEVs. On the basis of information

TABLE 5-3 Entities That Might Have an Incentive to Install Each Charging Infrastructure Category

Infrastructure Category		Who Has an Incentive to Install?
Location	Type	
Interstate	DC fast	Vehicle manufacturer, government
Intercity	DC fast	Vehicle manufacturer, government
Intracity	DC fast	Vehicle manufacturer, government, charging provider, utility
Intracity	AC level 1 or level 2	Utility, retailer, charging provider, vehicle manufacturer
Workplace	AC level 1 or level 2	Business owner, utility
Home or fleet base	AC level 1 or level 2	Vehicle owner, utility

NOTE: AC, alternating current; DC, direct current.

received during site visits and from presentations from various infrastructure providers, the committee's assessment of the possible builders of each infrastructure category is summarized in Table 5-3. It should be noted that the most critical infrastructure (home charging) is also the least logistically complicated and least expensive to build, and the costs and complications generally increase for faster charging and more public locations. The following paragraphs discuss the infrastructure-deployment models associated with each infrastructure segment and the installation and operating costs.

Home Charging

Private charging infrastructure at home is likely to be funded by the homeowner. Financing and logistics of installing home charging infrastructure is not considered to be an important barrier for homeowners who have dedicated parking spots adjacent to their homes. Homeowners who own PEVs have a clear incentive to install home charging. Many will also find the expense of upgrading to AC level 2 infrastructure to be a good investment, especially owners of long-range BEVs who might want to charge their vehicle batteries more quickly. Aside from vehicle owners paying to install charging infrastructure, other deployment models are being implemented. Some providers of subscription-based charging have expanded into providing residential charging infrastructure as part of their subscription service. Utilities might also have an interest in providing residential charging infrastructure as it would increase electricity usage at the residence.

As discussed previously, multifamily residential home charging faces many more barriers, and it is not clear that many owners of complexes, drivers of vehicles, or municipalities will have incentive to install charging at multifamily residences or at on-street charging locations in residential neighborhoods. However, owners of multifamily residences might be motivated to install chargers because they can earn points toward Leadership in Energy and Environmental Design certification (AeroVironment 2010). They might also be able to market their property as green and offer charging as an attractive amenity to prospective renters.

Workplace Charging

Private charging infrastructure at workplaces is likely to be funded by the businesses or organizations. The installation and operating costs of workplace charging might be justified by the employer as a perk to attract and retain employees or to brand the company with a green image. Because vehicles are parked at work for long periods of time (see Figure 5-2), many workplaces do not find it necessary to upgrade even to AC level 2 charging. Some parking lots might already have AC level 1 outlets that can be repurposed for vehicle charging; however, more convenient or upgraded infrastructure might also be installed. Another entity that might have an interest in installing workplace charging is a utility, which could earn additional revenue from the sale of electricity at worksites.

The cost of installing charging varies from workplace to workplace but is generally higher than that for installing single-family home charging and lower than that for public charging infrastructure. The costs of labor and conduit for installing charging units in existing parking lots and garages depend mainly on how much digging and resurfacing is involved. There are also potential costs associated with electric service upgrades for AC level 2 chargers, which might be the best choice for most currently available PEVs that have large electric ranges. Cisco provided a set of ballpark estimates to the committee and indicated that the average cost of installing an AC level 2 charging station has been $10,000-$15,000 (with economies of scale), that the ongoing costs of paying a vendor to manage the stations has averaged about $25 per station per month, and that the electricity costs have been low (Ahmed 2013). However, Bordon and Boske (2013) suggest that the cost of installing an AC level 2 charger in a commercial garage or on a public street ranges from $2,000 to $8,000 on the basis of estimates from three separate sources.

In addition to installation costs, operating costs of providing charging to employees must be considered. The committee received reports that the costs of electricity were not a barrier to deployment of workplace charging, but two logistical concerns were raised. As mentioned above and discussed further in Chapter 6, demand charges that could increase the cost of electricity to the employer could be a cost barrier to workplaces installing charging for employees. Also, the potential need to classify workplace charging as imputed income has resulted in logistical barriers given the associated administrative requirements for monitoring charging time or energy and making associated payroll adjustments. In part to avoid that potential problem and also to outsource charger installation and maintenance, some employers have chosen to contract with charging providers to install and operate charging infrastructure, including charging for the electricity provided.

Finding: Some workplaces appear to have incentives for installing charging infrastructure, including fostering an environmentally friendly image and providing the perk to retain and recruit employees.

Recommendation: Local governments should engage with and encourage workplaces to consider investments in charging infrastructure and provide information about best practices.

Public Charging Infrastructure

As discussed above, charging infrastructure generally becomes more complicated and more costly to build and operate as it becomes more publicly accessible and delivers faster charging. The potential owners and operators of public charging infrastructure are discussed in the sections below. Generally, companies that install and operate public charging stations have five sources from which they can seek to cover their capital and operating costs: the government, utilities, vehicle manufacturers, charging-station hosts, and drivers. Most companies have depended on government grants to finance a large part of their investments to date, and it is difficult to tell whether their business models will be sustainable in the absence of public funding.

The costs of DC fast-charging stations are generally much higher than the costs of AC level 2 stations. In general, the capital costs depend on several factors: whether the property must be purchased, leased, or rented; what distance must be spanned to connect to higher voltage supply lines; whether upgrades are required, for example, because of insufficient transformer capacity; how much trenching and conduit are needed to reach the charging station; and how much repaving or restriping of the parking area is required to accommodate the charging station. In total, the costs can range from \$100,000 to \$200,000. As an example, Table 5-4 shows the average costs of installing charging stations in Washington State with DC fast chargers and AC level 2 chargers as part of the publicly funded West Coast Electric Highway project. The totals shown in the table—ranging from \$109,500 to \$122,000—exclude the costs of purchasing, renting, or leasing land. The basic cost of a DC fast-charging station is about \$10,000 to \$15,000, but the total equipment cost of the Washington state stations averaged \$58,000, reflecting the auxiliary services and features needed for a publicly accessible unit, including warranty, maintenance, customer authentication, and networking with point-of-sale capabilities to collect payment from customers. Installation costs can also vary because of other enhanced safety and security measures that are often required by local permitting authorities, such as lighting and revenue-grade meters. Those options can add up to \$90,000 to the basic cost of the fast-charging equipment itself. Additional costs might also be incurred if multiple plugs are required for compatibility.

Retailers

A number of major retailers have shown interest in providing space for charging stations (Motavalli 2013),[6] particularly when the capital costs are subsidized. Such infrastructure can attract customers to park and spend time and money in the retail establishments and might also provide favorable branding for the retailers. Most of the charging units that retailers have provided to date have been AC level 1 or level 2 stations, which are used primarily for intracity charging. The costs of building charging infrastructure at retail establishments range widely but are probably similar to workplaces and related to the amount of conduit required to provide electric access at parking spots. It is not clear that the extra money spent in retail establishments by customers who use the charging stations is sufficient to provide retailers with incentives to incur the capital costs of installing charging stations, as distinct from simply covering electricity charges and service costs. When capital costs are covered by others, however, retailers have tended to contract with charging providers to build and maintain charging stations and possibly charge customers for their use.

Electric Utilities

The electric utility companies could emerge as a willing source of capital for public charging stations. That conclusion reflects the prospect that a network of public charging stations would induce more utility customers to purchase PEVs, which would lead not only to electricity consumption at the public chargers, but also to much greater consumption of electricity at residences served by the utilities. If public charging infrastructure drives greater eVMT and greater deployment of vehicles, capital and variable costs for public infrastructure might be covered by the incremental revenue from additional electricity that PEV drivers consume at home, where roughly 80 percent of PEV charging takes place (Francfort 2011). Most such charging infrastructure is expected to be built intracity. Austin Energy (2012), with the help of a series of federal government grants, is an example of a utility that has chosen

[6] Major retail companies that have installed or plan to install charging stations for their customers include Best Buy, Chili's, Cracker Barrel, Kroger, Macy's, 7-Eleven, Tim Hortons, Walgreens, and Whole Foods.

TABLE 5-4 Costs of Installing Public DC Fast-Charging Stations for the West Coast Electric Highway Project[a]

Component	Cost
DC fast-charging equipment • 50 kW DC public fast-charging station (480 V ac input) • 3-year warranty and point-of-sale capabilities[b] • Payment of all electricity dispensed (including utility demand charges) • Overhead lighting and required safety equipment	\$58,000 per unit
Level 2 charger colocated next to DC fast-charging station • 240 V/30 A AC level 2 public charger • Same terms and conditions as listed above	\$2,500 per unit
Equipment installation (labor and electric-panel upgrade) • Separate power drop or meter for the charging station • Electric panel upgrade (if required) • Construction and environmental and electricity permits • Trenching, backfill, and site restoration • Installation of conduit and power lines to charging station • Installation of concrete pad and electric stub-out • Installation of curb or wheel stop and overhead lighting • Installation and testing of equipment	\$26,000 per location
Utility interconnection • Costs are highly variable and depend on cost-recovery policies of the electric-power provider and condition of existing power distribution components[c] • Generally includes utility costs for preliminary engineering and design, transformer upgrades, and labor for connection to the grid	\$12,500 to \$25,000 per location
Host-site identification, analysis, and screening • Identification of potential sites • Consultation with electric-power providers	\$5,000 per location
Negotiation, legal review, and execution of lease • Making contact with several property owners • Exchanging and negotiating lease documents • Executing and recording documents	\$6,000 per location
Total for DC fast charger and 3-year service	\$109,500 to \$122,000

[a] Land costs are not included here.
[b] Point-of-sale capabilities might include radiofrequency identification authentication and networking to back-office functions (such as account management and customer billing), equipment status signals, and credit card transactions.
[c] Additional costs could be incurred if addition of multiple chargers increases demand charges or requires additional electricity service upgrades.
NOTE: A, amperes; AC, alternating current; DC, direct current; kW, kilowatt; V, volt.
SOURCE: Based on data from PB (2009).

to install a network of AC level 2 charging stations in its service area, where it is the only electricity provider, and to offer its residential customers unlimited use of the chargers for less than \$5 a month. In addition, Austin Energy provides incentives for a range of additional infrastructure charging categories as part of its strategic objectives in demand management and ancillary services (K. Popham, Austin Energy, personal communication, December 18, 2014).

The committee notes that theoretically all utilities servicing a given geographical area would collectively have a viable business model if there were a mechanism to (1) separate out and pool the electricity sales to all households that owned PEVs within that area and (2) share the revenues from that pool in proportion to the amounts that the different utilities contributed to investments in public charging infrastructure. Such a mechanism would not have to rely on government subsidies or cross subsidization from households that did not own a PEV. That said, whether utilities that invested their own capital in charging stations could earn a respectable rate of return over time would depend on state-level regulatory policies that are used to encourage utility investment.

Commercial Charging Providers

Mostly in response to government grants, several private companies have entered the business of installing and managing public charging stations. These charging stations are a mix of AC level 2 and DC fast chargers and are located both between and within cities. The companies have been experimenting with different models in their efforts to recover their capital costs and the costs of electricity. For example, ChargePoint (2014) is pricing on a per-charge-event basis, while NRG/eVgo (2014) relies on both a monthly subscription fee and a fee per minute of plug-in time. Depending on state legislative and regulatory rulings, charging providers might avoid being regulated as utilities by not charging in proportion to the amount of electricity consumed (see further discussion in Chapter 6). NRG/eVgo relies on its fee per minute of plug-in time as a mechanism for encouraging drivers to limit the amount of time that their vehicles occupy the parking spaces adjacent to the chargers.

Although it might be easy to cover the variable costs of their operations from the various fees paid by customers (for example, monthly subscription fees or fees per charging event or per minute of charging time), generating an attractive rate of return on invested capital is much more challenging. One of the early providers of charging infrastructure, ECOtality, encountered financial difficulties and filed for bankruptcy in October 2013; its Blink assets, including the network of Blink charging stations, have been purchased by CarCharging (Wald 2013).

The infrastructure-deployment model adopted by NRG/eVgo provides a unique approach. It is oriented toward providing a simple and complete set of services to residential customers. NRG/eVgo (2014) offers its Houston customers a 1-year contract for a $15 monthly fee that covers the installation of charging equipment at home and provides unlimited access to its network of public stations at 10 cents per minute of plug-in time. And unlike most other public stations, its Freedom Chargers include DC fast chargers and AC level 2 chargers and are located mainly along major transportation corridors within the metropolitan areas it serves.[7]

Box 5-1 provides a hypothetical calculation for the economics of providing public charging stations using a business model that collects monthly subscription fees and also charges customers for charging time. The calculation suggests that it might be difficult for charging providers to survive unless their capital costs are at least partially subsidized by public funding or by others, such as vehicle manufacturers. That said, the committee heard concerns from private charging developers that subsidizing infrastructure investments tended to undermine the business models of firms that were prepared to finance infrastructure with their own capital.

Vehicle Manufacturers

Vehicle manufacturers might deploy public charging infrastructure to drive sales of PEVs or to position themselves in the market. They might be one of the only private sector entities with a motive to install fast charging along intercity and interstate highways, as this type of infrastructure is the most expensive to build and is unlikely to generate high returns from for-pay charging. As noted earlier, Tesla has launched a program to install several hundred supercharging stations along major long-distance transportation corridors throughout the United States, while Nissan has launched several joint ventures to increase substantially the number of fast chargers available in key market areas (DeMorro 2014).

In the absence of government subsidies, it seems unlikely that any companies other than BEV manufacturers could have a business case for covering the installation and maintenance costs of DC fast-charging infrastructure deployed in intercity and interstate highway corridors. Whether the infrastructure would be publicly accessible is uncertain as a vehicle manufacturer would have little incentive for providing charging infrastructure for PEVs that it did not produce. For example, only Tesla customers can use Tesla-built chargers because of a Tesla-specific plug. In the case of Nissan, which is also building and subsidizing chargers, their chargers can be used by many types of PEVs but might require payment from those not covered under Nissan's No-Charge-to-Charge plan.

Federal Government

If a category of charging infrastructure is deemed to be particularly effective at inducing PEV deployment but no private sector entity has a strong case for building such infrastructure, the federal government might consider funding it as a worthwhile investment. The committee heard concerns that government money was likely to crowd out private investments in infrastructure and to lead to poor siting decisions in some cases. To ensure that charging infrastructure developers have an incentive to site chargers so that they will be well used, government infrastructure funding should comprise only a portion of the funding for a charging station and should not go toward stations that would be deployed without government funding. Also, more research should be done to ascertain what categories of charging infrastructure lead to increases in deployment and eVMT.

Finding: Utilities that can capture the entire residential electricity consumption of PEV owners appear to have a viable business model for investing in public charging infrastructure.

Finding: Initiatives undertaken by Tesla and Nissan suggest that vehicle manufacturers that wish to penetrate the market for BEVs perceive a business case for investing in extensive networks of DC fast-charging stations.

Finding: Apart from BEV manufacturers and utilities (or groups of utilities), the committee has not been able to identify any private sector entities that have an attractive business case for absorbing the full capital costs of investments in public charging infrastructure.

[7] eVgo areas include Houston, Dallas-Fort Worth, Los Angeles, San Francisco, San Diego, the San Joaquin Valley, and Washington, D.C. To the extent that the provision of a network of fast-charging stations helps catalyze PEV sales, total electricity consumption will increase by much more than electricity consumption at the eVgo charging stations and provide additional profits for NRG, which generates electricity.

BOX 5-1 Some Hypothetical Economics for Providers of Public Charging

This box considers the economics of providing a network of K public AC level 2 charging stations to serve N customers who rely primarily on residential charging but, on average, add 30 minutes of charge four times a month (or 1 hour of charge twice a month) at public chargers.

Assume that each charging station involves a capital outlay of $10,000 and that the investor requires a payback in 3 years, which in round terms amounts to about $3,600 per year per station, or $300 per month per station.

Assume that customers are charged 10 cents for each minute of plug-in time and that each hour of charging generates $2 of revenue over and above electricity costs plus maintenance costs. Thus, use of the charging network generates net revenues of $\$4N$ per month.

Assume that customers are willing to pay a subscription fee of $\$F$ per month for the assurance of access to the network of stations, implying subscription revenue of $\$NF$ per month.

Then the break-even value of N, calculated as a function of F, must satisfy $NF = 300K - 4N$, or

$$N = 300K/(F + 4)$$

And the break-even value of F as a function of K/N can be expressed as

$$F = 300(K/N) - 4$$

This suggests that a firm with 200 subscribers for every 10 charging stations could break even by charging a subscription fee of $11 per month.

Note, however, that the economics becomes much more difficult for networks of DC fast chargers, which require much larger capital outlays, or for AC level 2 networks that have to compete with networks of fast chargers.

Finding: The federal government might decide that providing public charging infrastructure serves a public good when others do not have a business case or other incentive to do so.

Recommendation: The federal government should refrain from additional direct investment in the installation of public charging infrastructure pending an evaluation of the relationship between the availability of public charging and PEV adoption or use.

REFERENCES

AeroVironment. 2010. "EVs and LEED Certification." AV Connect, October. http://www.avinc.com/plugin/nl002/evs_and_leed_certification.

Ahmed, A. 2013. "Workplace Electric Vehicle Charging." Presentation to the Committee on Overcoming Barriers to Electric-Vehicle Deployment, Washington, DC, August 13.

Andrews, M., M.K. Dogru, J.D. Hobby, Y. Jin, and G.H. Tucci. 2013. "Modeling and Optimization for Electric Vehicle Charging Infrastructure." Innovative Smart Grid Technologies Conference, Washington, DC, February 24-27.

Anegawa, T. 2010. "Development of Quick Charging System for Electric Vehicle." World Energy Congress, Montreal, Canada, September 12-16. http://89.206.150.89/documents/congresspapers/322.pdf.

Austin Energy. 2012. "For Less Than $5 Per Month, Austin Energy Offers Unlimited Electric Vehicle Charging at More Than 100 EVerywhere™ Stations." Press release, March 2.

Borden, E.J., and L.B. Boske. 2013. Electric Vehicles and Public Charging Infrastructure: Impediments and Opportunities for Success in the United States. Center for Transportation Research, University of Texas at Austin, July.

CarCharging. 2013. "Nissan and CarCharging to Expand Electric Vehicle Quick Charger Network." News release, May 14. http://www.carcharging.com/about/news/all/nissan-and-carcharging-to-expand-electric-vehicle-quick-charger-network/.

ChargePoint. 2014. "Start Charging Today." https://www.chargepoint.com/get-started/. Accessed October 23, 2014.

Chen, T.D., K.M. Kockelman, and M. Khan. 2013. "The Electric Vehicle Charging Station Location Problem: A Parking-Based Assignment Method for Seattle." Presented at the 92nd Annual Meeting of the Transportation Research Board, Washington, DC, January 13-17. http://www.caee.utexas.edu/prof/kockelman/public_html/TRB13EVparking.pdf.

Consumers Union and the Union of Concerned Scientists. 2013. "Electric Vehicle Survey Methodology and Assumptions: American Driving Habits, Vehicle Needs, and

Attitudes Towards Electric Vehicles." December. http://www.ucsusa.org/assets/documents/clean_vehicles/UCS-and-CU-Electric-Vehicle-Survey-Methodology.pdf.

DeMorro, C. 2014. "Nissan Launches "No Charge to Charge" Program for LEAF Buyers." Clean Technica, July 10. http://cleantechnica.com/2014/07/10/nissan-launches-charge-charge-program-leaf-buyers.

DOE (U.S. Department of Energy). 2013. Plug-In Electric Vehicle Handbook for Workplace Charging Hosts. Clean Cities, August. http://www.afdc.energy.gov/uploads/publication/pev_workplace_charging_hosts.pdf.

DOE. 2014a. "California Laws and Incentives." Alternative Fuels Data Center. Last updated November 2014. http://www.afdc.energy.gov/laws/state_summary?state=CA.

DOE. 2014b. "EV Everywhere Workplace Charging Challenge." http://energy.gov/eere/vehicles/ev-everywhere-workplace-charging-challenge. Accessed December 30, 2014.

DOE. 2015. "Alternative Fuels Data Center: Alternative Fuels Station Locator." http://www.afdc.energy.gov/locator/stations. Accessed April 8, 2015.

ECOtality. 2013. "What Kind of Charging Infrastructure Do Chevrolet Volt Drivers in The EV Project Use?" The EV Project. Lessons Learned White Paper. September. http://avt.inl.gov/pdf/EVProj/VoltChargingInfrastructureUsageSep2013.pdf.

ECOtality. 2014a. "Where Do Nissan Leaf Drivers in The EV Project Charge When They Have the Opportunity to Charge at Work?" The EV Project. Lessons Learned White Paper. March. http://avt.inel.gov/pdf/EVProj/ChargingLocation-WorkplaceLeafsMar2014.pdf.

ECOtality. 2014b. "Where Do Chevrolet Volt Drivers in The EV Project Charge When They Have the Opportunity to Charge at Work?" The EV Project. Lessons Learned White Paper. March. http://avt.inel.gov/pdf/EVProj/ChargingLocation-WorkplaceVoltsMar2014.pdf.

Francfort, J. 2011. "U.S. Department of Energy's Vehicle Technology Program: Plug-in Electric Vehicle Real World Data from DOE's AVTA. Idaho National Laboratory, EPRI Infrastructure Working Council, Tempe, AZ.:" December. http://mydocs.epri.com/docs/publicmeetingmaterials/1112/PDNNSH5Q3Z2/Day%201%20PlugIn%20Electric%20Vehicle%20Codes%20&%20Standards.pdf.

INL (Idaho National Laboratory). 2014. EV Project Electric Vehicle Charging Infrastructure Summary Report: October 2013 through December 2013. INL/MIS-10-19479. http://avt.inel.gov/pdf/EVProj/EVProjectInfrastructureQ42013.pdf.

IRS (Internal Revenue Service). 2014. Publication 15-B: Employer's Tax Guide to Fringe Benefits. Washington, DC: Government Printing Office.

Jennings, D. 2013. "EV Charging at Lynda.com." Presentation to the Committee on Overcoming Barriers to Electric-Vehicle Deployment, Washington, DC, August 13.

Motavalli, J. 2013. "Shop While You Charge: Consumers and Retailers Are Plugging In." New York Times, July 9.

Musgrove, R. 2013. "Global Vehicles." Presentation to the Committee on Overcoming Barriers to Electric-Vehicle Deployment, Washington, DC, December 3.

Nicholas, M., G. Tal, and J. Woodjack. 2013. "California Statewide Charging Assessment Model for Plug-in Electric Vehicles: Learning from Statewide Travel Surveys." Working Paper UCD-ITS-WP-13-01. Institute of Transportation Studies, University of California, Davis. January.

Nissan. 2013. "Nissan to Triple Electric Vehicle Fast-Chargers; Partners with NRG Energy to Build First Fast-Charge Network in Washington D.C. Area." Press release, January 31. http://nissannews.com/en-US/nissan/usa/releases/nissan-to-triple-electric-vehicle-fast-chargers-partners-with-nrg-energy-to-build-first-fast-charge-network-in-washington-d-c-area.

Nissan. 2014. "Nissan launches programs to make LEAF charging free and 'EZ.' Press release, April 16. http://nissannews.com/en-US/nissan/usa/releases/nissan-launches-programs-to-make-leaf-charging-free-and-ez.

NRG/eVgo. 2014. "Electric Car Charging Plans." http://www.nrgevgo.com.

PB (Parsons Brinckerhoff). 2009. "Alternative Fuels Corridor Economic Feasibility Study." Prepared for Washington State Department of Transportation, Office of Public/Private Partnerships, January 23. http://www.wsdot.wa.gov/NR/rdonlyres/5C14E610-713A-4600-A88D-C567AF49D096/0/AltFuelsFinalReport.pdf.

Peterson, D. 2011. Addressing Challenges to Electric Vehicle Charging in Multifamily Residential Buildings. The Luskin Center for Innovation. Los Angeles, CA. June. http://www.luskin.ucla.edu/sites/default/files/EV_Multifamily_Report_10_2011.pdf.

Peterson, S., and J. Michalek. 2013. Cost effectiveness of plug-in hybrid electric vehicle battery capacity and charging infrastructure investment for reducing U.S. gasoline consumption. Energy Policy 52: 429-438.

Recargo. 2014. "Charging Station Finder." http://www.plugshare.com/. Accessed October 23, 2014.

Smart, J. 2014a. "PEV Infrastructure Deployment Costs and Drivers' Charging Preferences in the EV Project." SAE 2014 Hybrid and Electric Vehicle Technologies Symposium, La Jolla, CA, February 11. http://avt.inl.gov/pdf/EVProj/SAEHybridEVSympFeb2014.pdf.

Smart, J. 2014b. "Workplace Lessons Learned through the Nation's Largest PEV Charging Projects." DOE Workplace Charging Challenge Summit, Alexandria, VA, November 18. http://avt.inl.gov/pdf/EVProj/WorkplaceChargingDataInsights.pdf.

Smart, J. 2014c. "EV Charging Infrastructure Usage in Large-Scale Charging Infrastructure Demonstrations: Public Charging Station Case Studies for ARB." Plug-in Electric Vehicle Infrastructure Information Gathering Meeting,

July 15. http://avt.inel.gov/pdf/EVProj/EVInfrastructure-UsageARBJul2014.pdf.

Smart, J. 2014d. "Latest Insights from The EV Project and ChargePoint America PEV Infrastructure Demos." Presentation at the GITT meeting at INL, August 12. http://avt.inel.gov/pdf/EVProj/GITTEVProjectAug2014.pdf.

Smart, J., and S. Schey. 2012. Battery electric vehicle driving and charging behavior observed early in the EV Project. SAE International Journal of Alternative Powertrains 1(1): 37-33.

Smart, J., and S. White. 2014. "Electric Vehicle Charging Infrastructure Usage Observed in Large-scale Charging Infrastructure Demonstrations." Presentation to the Committee on Overcoming Barriers to Electric-Vehicle Deployment, Irvine, CA, February 25.

Tate, E., and P. Savagian. 2009. "The CO_2 Benefits of Electrification E-REVs, PHEVs and Charging Scenarios." SAE Technical Paper 2009-01-1311. SAE International, Warrendale, PA.

Tesla. 2014. "Supercharger: The Fastest Charging Station on the Planet." http://www.teslamotors.com/supercharger. Accessed May 18, 2014.

Traut, E.J., C.C. TsuWei, C. Hendrickson, and J.J. Michalek. 2013. US residential charging potential for electric vehicles. Transportation Research Part D 25: 139-145.

U.S. Census Bureau. 2011a. "2011 American Housing Survey for the United States, Table C-02-AH." http://www.census.gov/housing/ahs/data/national.html.

U.S. Census Bureau. 2011b. "2011 American Housing Survey for the United States, Table C-01-AH." http://www.census.gov/housing/ahs/data/national.html.

Vance, A. 2013. "Gone in 90 Seconds: Tesla's Battery-Swapping Magic." Bloomberg Businessweek, June 21. http://www.businessweek.com/articles/2013-06-21/gone-in-90-seconds-teslas-battery-swapping-magic.

Wald, M. 2013. "A Recharging Industry Rises." New York Times, November 12. http://www.nytimes.com/2013/11/13/business/energy-environment/electric-cars-give-rise-to-a-recharging-industry.html.

6

Implications of Plug-in Electric Vehicles for the Electricity Sector

An important component of the ecosystem of the plug-in electric vehicle (PEV) is the electric utility, which provides the electricity that powers the vehicle.[1] Electric utilities in the twenty-first century have experienced eroding demand (see Figure 6-1) and view PEVs as a potential source of increased demand (Kind 2013; EEI 2014). The Edison Electric Institute, the largest trade association for electric utilities, contends that the industry needs increased electrification of the transportation sector for the electricity sector to remain viable and sustainable in the long term (EEI 2014).

An important concern raised by the public and policy makers, however, is the ability of electric utilities to accommodate PEV charging, a concern that impacts not only PEV owners but also the public more broadly. At the current time, PEV charging requirements account for about 0.02 percent of the energy produced and consumed in the continental United States (EIA 2012).[2] Were the share of the PEV fleet to reach as high as 20 percent of private vehicles, the estimated impact would still account for only 5 percent of today's electricity production (DOT 2014; EIA 2012).[3] Accordingly, the electricity sector does not perceive PEVs as posing any near-term or mid-term challenges. However, some have assumed that electric utilities cannot accommodate transportation electrification with the current grid infrastructure. That mistaken belief is also held in other countries and has been cited as a key reason why electric utilities have not been allowed to take a more proactive role in facilitating the deployment of PEVs and the associated charging infrastructure (Anegawa 2010). Therefore, it is important to examine the current electricity sector and consider what impediments might exist.

Accordingly, this chapter examines potential impediments from the perspective of the individual components of electric utilities (the distribution, transmission, and generation components) and overall system control. To put the discussion in context, the committee first describes the physical and economic structure of electric utilities. Physical constraints in the distribution infrastructure for PEV charging are identified next, followed by a discussion of potential economic constraints and impediments within the delivery system. One scenario for a hypothetical utility of the future is described at the conclusion of the chapter. The committee's findings and recommendations are provided throughout the chapter.

One important point that should be noted before beginning the discussion of the electricity sector is that the federal government has only limited powers in directly influencing or modifying the policies and behavior of the owners or operators of the retail electricity sector. Although the Federal Energy Regulatory Commission (FERC) maintains authority to regulate transmission and wholesale sales of energy in interstate commerce, the retail electricity sector is regulated heavily and almost entirely by individual state regulatory commissions. Thus, the ability of private-investor-owned electric utilities to foster or impede the development of PEVs will vary significantly based on the actions of the individual state utility commissions. Furthermore, different regulatory bodies oversee municipal-owned utilities, federally owned utilities, cooperative utilities, and, as indicated, the wholesale markets. These jurisdictional and regional regulatory differences limit the federal government's ability to affect the practices of the U.S. electricity sector (see, for example, U.S. Court of Appeals 2014 decision on FERC Order 745).

Finding: State jurisdiction over retail electric rates constrains the federal role in directing the electricity sector to foster PEV growth.

---

[1] An electric utility is a publicly or privately owned company that generates, transmits, and distributes electricity for sale to the public and includes vertically integrated utilities that own their generation plants, transmission components, and distribution wires and unbundled utilities that separate the generation, transmission, distribution, and retail into different businesses. Although the majority of electric utilities in the United States are privately owned, there are a substantial number of generally smaller utilities that are owned and operated by regional organizations or municipal governments, often referred to as *munis*. The largest muni in the United States is the Los Angeles Department of Water and Power.

[2] This estimate assumes that each PEV consumes about 10 kWh/day.

[3] This estimate assumes the aforementioned consumption for vehicle charging and that there would be 192.5 million light-duty vehicles on the road, which is equivalent to the number in 2011 in the United States (DOT 2014).

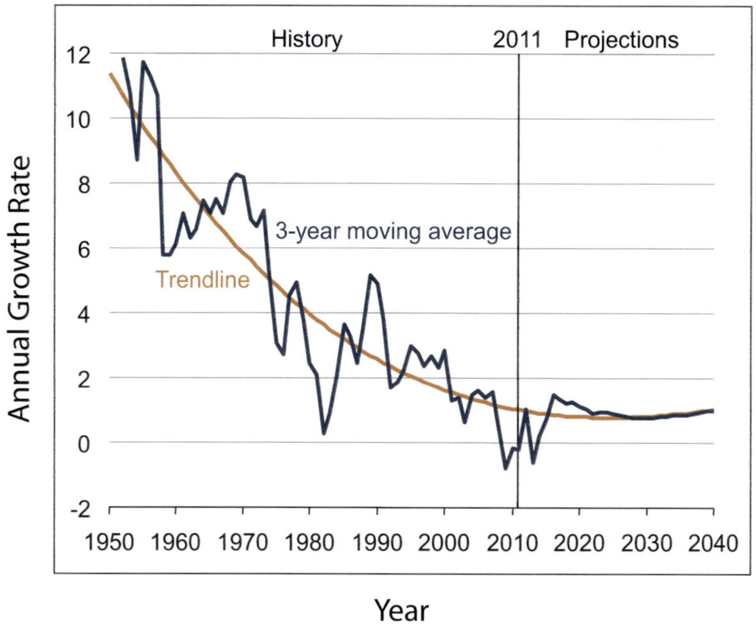

FIGURE 6-1 U.S. electricity demand growth, 1950-2040. From the 7 percent annual growth rates from the 1950s through the 1970s to the declines of the 1980s and 1990s when average growth in demand was about 3 percent per year, the first decade of this century has been nearly flat with an average growth rate of only 0.7 percent. SOURCE: EIA (2013).

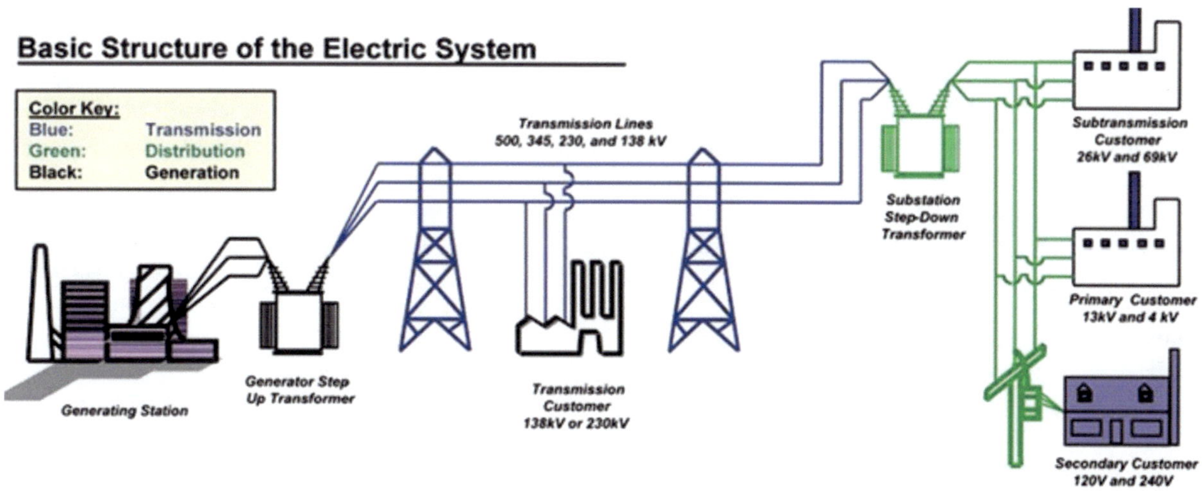

FIGURE 6-2 Schematic of U.S. electric power delivery system. SOURCE: U.S.-Canada Power System Outage Taskforce (2004).

THE PHYSICAL AND ECONOMIC STRUCTURE OF THE ELECTRICITY SECTOR

Figure 6-2 is a schematic of the U.S. electricity sector. Generation companies produce electricity from fossil or non-fossil (nuclear and renewable) sources. Transmission entities are responsible for high-voltage transmission and frequently for overall system control. Distribution companies are publicly or privately owned companies that sell, state by state, price-regulated electric energy to retail customers, residential, commercial, and industrial. They may be independent or part of a vertically integrated electric utility.

Today's structure of the electricity sector and the business entities within it have been in a state of constant flux and evolution since April 1996, when the FERC issued its Orders No. 888 and 889, which formally separated generation, transmission, and distribution from each other, thereby providing open access to transmission in the United States to any generating entity and allowing for the operation of highly fluid wholesale electric markets (FERC 1996). In the Northeast, Midwest, Southwest, Texas (ERCOT), and California, Order 888 has resulted in the creation of Independent System Operators (ISOs); in the remainder of the country, Regional Transmission Operators (RTOs) act as wide area system operators. The ISOs and RTOs operate and control the transmission system and manage the organized wholesale markets between generators and retail suppliers and large industrial customers. Independently owned electricity generators operate by selling wholesale electricity into organized or bilateral markets; that electricity is transmitted by separate corporate and operational entities to distribution companies, which serve retail consumers.

GENERATION AND TRANSMISSION

For roughly 60 percent of the United States, the electricity sector operates through organized markets coordinated by ISOs (EIA 2011). Most electric consumers in the United States get their energy from generators within large, centrally controlled regional networks. Their energy is transmitted over high-voltage wires that are regulated by the FERC. That energy is finally delivered through a distribution system regulated by state public utility commissions (PUCs) that are responsible for setting the price paid per kilowatt hour. Where states have opted for retail competition, such as in Ohio and Texas, the state commissions oversee and approve the manner in which the sellers of retail energy structure their services rather than set the price per kilowatt hour for electricity delivered to consumers.

Understanding the electric power delivery chain is critical for understanding the current and future interactions between electric utilities and PEV charging systems and for identifying any impediments that might be introduced by electric utilities. As with virtually all end uses of electricity, the point of contact between the electricity sector and the end user is the distribution company, regardless of whether it is residential charging, public charging, or fleet charging. It is at the local electricity distribution level that concentrations of PEVs might stress the delivery infrastructure (Maitra 2011). However, even with high adoption rates for PEVs and therefore for vehicle charging, the impact on the electricity system at large is insignificant.

Although both the generation and the transmission sectors are critical to the ultimate delivery of electricity for vehicle charging, they are not an impediment to PEV acceptance because meeting the demand created by PEV charging is well within the planning and operational capability of the electricity sector. From the perspective of the largely competitive wholesale electricity market, any increase in demand is welcome, particularly demand that has the potential to smooth daily variability (a characteristic of vehicle charging).

Finding: There is no anticipated impact on either the generation or the transmission sector of the U.S. electric power system from the introduction of PEVs. Thus, the existing capability to generate and transmit power within the United States is not now nor is it anticipated to be a deterrent to the adoption of PEVs.

PHYSICAL CONSTRAINTS IN THE DISTRIBUTION INFRASTRUCTURE

Although the introduction of PEVs is not constrained by the transmission system or the generation capacity, the electric sector distribution infrastructure, which is a lower voltage and lower capacity segment of the electric power system, could face operational constraints. PEVs are not, nor are they anticipated to be, uniformly distributed within the country or any region but are instead generally expected to be locally concentrated (see Chapter 3). PEVs have typically been concentrated in specific geographic areas that have higher median incomes, place higher values on environmental issues and energy security, and have higher average educational levels. Those demographics suggest that PEV acquisition will be concentrated in particular residential areas of the distribution system. As a result, any of the potential problems for the distribution system noted above will most likely be localized (Maitra 2011). Several scenarios in which problems could arise are discussed below.

The first scenario in which PEVs could pose an operational constraint on the distribution infrastructure is when several PEVs are simultaneously being charged on one transformer or one branch circuit that was designed to serve the traditional loads of a few residences. In that scenario, PEV charging could affect power system stability; for example, charging could cause a voltage drop in the local distribution system or cause voltage and current phase imbalances. Thus, the introduction of several PEVs could necessitate upgrades to the distribution system, such as a new transformer or a larger branch circuit that would not otherwise have been needed.

The charging of an individual PEV could be a challenge to the distribution company if that charging is coincident with peak electricity consumption on any individual distribution system element operating at full capacity. It would be extremely rare for PEV charging to coincide in time with the distribution company's peak, which typically occurs between noon and 6 p.m. It is more likely that a PEV would be charging at a time that coincides with the peak electricity usage of a residential circuit, which is typically between 5 p.m. and 11 p.m. That scenario at the residential circuit level could overload four components of the distribution infrastructure: the service drop (the wire from local transformer to the home or other point of charge), the local distribution transformer, feeders (wires from local distribution transformer to distribution substation), or a substation transformer. Figure 6-3 provides an example of hourly demand for electricity at a substation within a residential distribution system and illustrates the pattern of residential consumption for several cases. Case 1 illustrates what might happen without any incentives for off-peak charging. It shows a measurable impact on the peak and indicates that without incentives to reduce charging on peak, there could be specific locations where additional capital investments might be needed to accommodate the added demand from PEV charging.

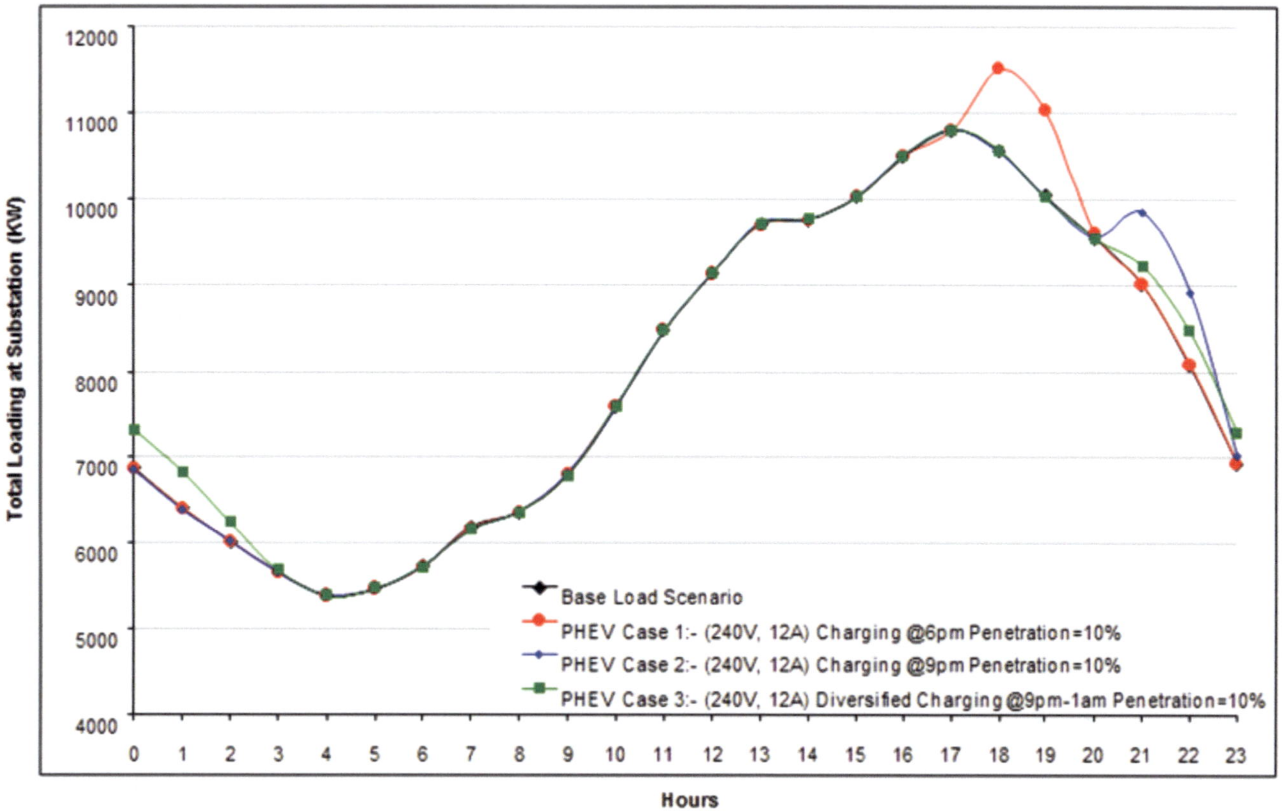

FIGURE 6-3 Hourly demand for electricity at a substation in a residential distribution system. NOTE: A, amperes; kW, kilowatt; PHEV, plug-in hybrid electric vehicle; V, volt. SOURCE: Maitra et al. (2009). Image courtesy of Electric Power Research Institute.

From the perspective of the distribution company, PEV charging represents an added uncertainty for the planning process. There are multiple dimensions to the issue, including how many PEVs will be purchased, where PEVs will be charged, and whether the pattern of charging will be coincident with local peak electricity consumption. Finally, there is the question of whether there are state-regulator-approved actions that the distribution company can take to alter the pattern of charging demand to minimize or potentially eliminate any negative effects, such as strong pricing incentives, timing restrictions, or indirect or direct charging control. Research done in California on different pricing incentives shows that PEV owners are price responsive, that larger price differentials encourage customers to charge off-peak, and that customers tend to remain on these time-of-day, price differentiated tariffs (CPUC 2012a).

Research also indicates that even without time-differentiated rates, PEV charging patterns tend to follow a pattern that has only moderate effects on distribution system peaks (CPUC 2014). With the near-term adoption levels anticipated for PEVs, there is still a natural diversity in the time and scale of PEV charging that is dictated by the type of trips that are taken in the vehicles. Currently, PEV charging behavior exhibits a gradual load curve that peaks at about 7 p.m., when most PEV owners arrive at home from work and plug in to charge at the same time. Even then, the number of upgrades at the distribution level has been minor—less than 0.75 percent of PEVs have required a local distribution system to upgrade a component—and has cost ratepayers only \$36,029 overall (CPUC 2014).

Another study, by the largest California utilities (E3 2014), demonstrates that even at high PEV adoption levels, the impacts on the distribution grid are minimal. The E3 study used the distribution data and load patterns for the California utilities, analyzed the distribution of PEV adoption at the 9-digit zip code level, and forecast the incremental cost from PEVs on each individual distribution line and transformer until 2030 for two scenarios: a normal case that meets the California zero-vehicle-emission mandate and a case that has adoption levels three times higher than the normal case. The study found that even for the highest adoption levels, the cost would be less than 1 percent of the annual distribution-upgrade costs of the California utilities. The study also found that time-of-use charging would reduce costs to customers by 60 percent compared with charging at any time during the day.

Some concern has been expressed about future patterns of charging and the resulting impact on the reliability of the distribution system with the introduction of DC fast charging (see Table 5-1). However, because the typical driving distance for a PEV is not likely to change because of fast charging, the higher charging levels simply mean that PEVs will charge in a shorter period of time while requiring the same overall quantity of energy. The higher power, shorter duration charging is unlikely to have a substantial effect on the distribution infrastructure. Furthermore, data from the EV Project indicate that DC fast charging represents only a small proportion of charging for vehicles (less than 1 percent of the energy demand for the Nissan Leafs in the study) (INL 2014).

Finding: PEV charging has had a negligible effect on the distribution-system components to date and is expected to have a negligible future effect at the anticipated rates of PEV adoption.

POTENTIAL ECONOMIC CONSTRAINTS OR IMPEDIMENTS WITHIN THE DELIVERY SYSTEM

With its existing capabilities, the generation and transmission elements of the U.S. electric power system are sufficiently robust to provide the infrastructure and deliver the energy required for PEV charging. As indicated above, any physical constraints or impediments to the distribution system will be highly localized and most likely will be only within individual distribution branches in the near to midterm. Thus, any constraints on PEV adoption that could arise from the electricity sector are more likely to be economic rather than physical or technical.

The economic constraints are primarily associated with two factors: high underlying electricity costs and ineffectively aligned rate structures. High underlying electricity costs reduce the financial benefit of owning a PEV by making the costs to drive the PEV closer to those of an ICE vehicle. The electricity cost is most often a function of the underlying characteristics of generation on a regional basis, with the hydroelectric generation of the Northwest producing much less expensive electricity than fossil-fuel generation of the Northeast. The regional differences in electricity costs add confusion to uniform explanations of the economic operating benefits of PEV ownership, as noted in Chapter 3.

A minor economic concern is the small possibility that system upgrades could in some cases be charged directly to the PEV-owning customers who necessitate the upgrade. If that cost were charged to an individual or small set of customers, it would substantially raise their costs of owning and operating a PEV. The handling of any cost allocation would depend on distribution company tariffs that govern whether individual customers are responsible for any electricity system upgrades that are incurred solely on their behalf or whether those costs can be spread over all electric customers.

The distribution company rate tariffs that are offered to end-use retail customers could raise obstacles to PEV adoption, including (1) inconsistency between rate tariffs, (2) lack of price incentives, (3) high average costs for electricity usage for residential customers, and (4) high costs for commercial and industrial customers due to demand charges (see Table 6-1 for descriptions of various rate structures). These potential obstacles can confuse retail customers about the best available electricity rate and the price advantage that they might receive by using electricity as a transportation fuel. Commercial consumers might have the added disincentive of a demand charge that is triggered by increased peak load.

The price paid by the end user for energy varies substantially between customer classes—industrial, commercial, and residential—and varies even more substantially from region to region, state to state, and distribution company to distribution company. State-regulated rate structures are designed to allow a regulated retailer to recover its fixed and variable costs and earn a fair rate of return. The costs include the variable cost of generated or purchased energy and a return on capital invested in generation, transmission, and distribution along with the operating costs of the company. The task of the PUCs is to allocate the full and reasonable costs of providing reliable energy across time, geography, and customer class. State jurisdictional authority in setting retail electricity rates has resulted in little or no consistency in the final price of electricity in terms of both the absolute price per kilowatt-hour of electricity and the rate structure itself. Uniform change appears to be nearly impossible given the fact that electric tariffs seen by all consumers (residential, commercial, and industrial) vary widely as a function of the underlying energy generation structure, the tax structures that the distribution companies face, and the vagaries of being regulated by 50 different state regulators and the local regulatory bodies that oversee more than 2,000 municipal and cooperative utilities. On the other hand, that same variability has allowed for multiple experiments in how to design rate structures for PEV charging.

The substantial differences in electric rates from one utility to another and between states are impediments to PEV adoption because it prevents a sales campaign from communicating easily or simply the economic benefits and costs of PEVs to potential buyers. Consumers have become accustomed to translating mpg values in national advertising for ICE vehicles, recognizing that the price of gasoline varies by at most 10 to 20 percent across the country. Compare that with the variability in the residential cost of electricity between Connecticut (18.22 cents per kilowatt-hour) and Washington State (8.7 cents per kilowatt-hour), with the former slightly more than double (EIA 2014). That spread does not account for any differentials in peak and off-peak rates, if they exist, or any demand charges that might be applied. Also, it does not consider the variety of types of PEVs,

TABLE 6-1 Definitions, Advantages, and Disadvantages of Various Types of Electric Rates

Type of Rate	Relevant Definition	Advantages	Disadvantages
Flat rate	An average rate charged volumetrically in cents per kWh; it would apply to all usage (e.g., \$0.18/kWh).	▪ Simple and understandable.	▪ Unlikely to reflect cost causation fully (the impact of a customer's consumption on the total cost of the system). ▪ Does not encourage behavioral changes in energy usage (e.g., switching from on-peak to off-peak).
Fixed charge and volumetric charge	Fixed charges are monthly charges (e.g., \$5/month) applicable to all customers regardless of usage; they are intended to reflect costs that do not change with usage and are necessary to ensure constant availability of service. Volumetric charges are per kWh charges based on electricity usage during the billing cycle (e.g., \$0.15/kWh); they are intended to reflect costs that change with usage (e.g., variable generation charges) and typically include generation, distribution, transmission, and public purpose program costs.	▪ Simple and understandable. ▪ May better reflect cost-causation. ▪ Reflects the per customer fixed costs required to serve each customer on per month basis.	▪ Likely to increase bills for low-use customers compared with flat rate or tiered structure. ▪ Fixed charges might not fully reflect cost-causation for classes of customers (e.g., multifamily vs single-family residences). ▪ Might decrease incentives to conserve.
Tiered rates	A rate that changes as a function of cumulative customer electricity usage in a monthly bill cycle. The tiers generally are defined from a baseline quantity or monthly minimum. Prices in an "inverted tier" or "inclining block" rate increase as cumulative electricity usage increases. For example, Tier 1, electricity usage up to the baseline amount; Tier 2, electricity usage from 101 to 130 percent of baseline; Tier 3, electricity usage from 131 to 200 percent of baseline; Tier 4, electricity usage greater than 200 percent of baseline.	▪ Baseline promotes affordability for basic needs. ▪ Higher tier rates are perceived to encourage conservation. ▪ Two tiers are relatively simple to understand compared with a more complex rate structure.	▪ Does not reflect different consumption needs of single vs multifamily residences. ▪ Multiple tiers are more difficult for customers to understand. ▪ Increased monthly usage might not actually raise utility's cost per kWh. ▪ Price signals are apparent only after bill is received. ▪ Poorly differentiated tiers might cause significant cross subsidies between customer groups.
Demand charges	Calculated on a per-kW basis for a customer's monthly maximum power usage (e.g., \$5/kW). Demand charges are generally calculated to reflect the cost of transmission and distribution facilities built to meet customers' maximum power demands. Demand charges are in addition to volumetric energy charges (per kWh), but the volumetric energy charges are lower than those on rate schedules without demand charges.	▪ Might better reflect cost causation.	▪ Might not be simple and understandable for residential customers (typically used for larger, more sophisticated commercial and industrial customers). ▪ Likely to increase bills for low-use customers compared with tiered structure. ▪ Could discourage energy efficiency, conservation measures, customer demand response, and customer generation.
Time-of-Use (TOU) rate	A rate that prices electricity according to the season or time of day that it is used. A TOU rate design more closely reflects the actual cost of providing electricity; it is characterized by ▪ Lower rates during a utility's off-peak and partial-peak demand periods. ▪ Higher rates during seasonal and daily peak demand periods. ▪ Because TOU rates are higher during the peak period, when incremental costs are highest, they send more accurate price signals to customers.	▪ Accomplishes several goals: reflects economic value (marginal cost) of energy, encourages conservation and reduces peak use, and leads to economically efficient decision making. ▪ Depending on their consumption pattern, customers could see lower or higher bills. ▪ Encourages off-peak charging of PEVs and greater use of residential or commercial solar photovoltaics.	▪ Could cause some customers' bills to increase, especially those with above-average peak-period usage. ▪ Might increase carbon emissions if coal is the marginal generating fuel during low-priced, off-peak hours of charging.
Critical Peak Pricing (CPP)	A dynamic rate that allows a short-term price increase to a predetermined level (or levels) to reflect real-time system conditions. In a fixed-period CPP, the time and duration of the price increase are predetermined, but the days are not predetermined. CPP programs provide participating customers an incentive to shift usage away from peak hours on a CPP event day. CPP event days are generally called 24 hours in advance.	▪ Provides direct peak reduction for the utility. ▪ Enrolled customers who respond to event notifications will see lower bills.	▪ Enrolled customers that do not respond to event notifications are likely to see bill increases.
Dynamic rate	A dynamic rate allows prices to be adjusted at short notice (typically an hour or a day ahead) as a function of system conditions. Either the price or the timing or both are unknown until real-time system conditions warrant a price adjustment. Examples include real-time pricing (RTP) and critical peak pricing (CPP). RTP allows prices to be adjusted frequently, typically on an hourly basis, to reflect real-time system conditions.	▪ Accomplishes several goals: reflects economic value (marginal cost) of energy, encourages conservation and reduces peak use, and leads to economically efficient decision making.	▪ Customer response to RTP is enhanced by and generally requires technology controls for monitoring and response. ▪ Might increase carbon emissions if coal is the marginal generating fuel during low-priced, off-peak hours of charging.

SOURCE: Based on data from CPUC (2012b).

which include BEVs that run only on electricity and PHEVs that can run on a gasoline or electricity, and whose mix of those fuels will vary by battery capacity and driving needs. Assembling a broad message for consumers on costs and benefits is practically impossible given that fuel costs vary, on average, by a factor of at least two and can vary by a factor of 4 or more.[4,5] The difficulty in generalizing fueling costs is discussed further in Chapter 3.

Residential electric rate structures for vehicle charging can also be an impediment to PEV adoption. Flat rates provide no incentive for the owner to charge the vehicle at the optimal time for the utility. Given that flat rates represent averages over a broad customer base, if the PEV is used for commuting and thus is charged at night, for example, the flat rate is likely to be high relative to the distribution company's actual marginal cost of supplying electricity at that time and at that location within the distribution system. The incentives provided by time-of-use (TOU) rates are substantially better aligned with the true costs of serving electric customers but add to the distribution company's cost if digital, multiregister meters are not already installed at the home. Although time-differentiated rates generally benefit PEV owners, they can be a disincentive if owners need to charge during high-priced, generally midday, time periods. Figure 6-4 shows an example of the impact of TOU rates on charging behavior as reported in the EV Project (ECOtality 2013). The time during which a vehicle was connected to a residential charger and the time during which the vehicle was actually drawing power were examined in the service territories of the Nashville Electric Service (NES) and Pacific Gas and Electric (PG&E). NES does not offer TOU rates, but PG&E does. Figure 6-4 shows that while the vehicles were connected to residential chargers for similar times in the two service areas, demand for charging energy was very different in the service area with TOU pricing, PG&E. That finding indicates that user behavior in plugging in the vehicle is the same for both regions but that TOU pricing motivates customers to use the timers integrated with the vehicle or charger to control their charging time and minimize their cost. Given that PEV charging at residential sites most often is discretionary, in that it can occur any time after the vehicle returns home and before it is needed the next day, PEV owners can take advantage of time-differentiated rates to charge their vehicles during the least costly period, benefiting both the owner and the utility.

TOU pricing has been in place at Tokyo Electric Power Company (TEPCO) for over 20 years (TEPCO, personal communication, December 10, 2013). TEPCO says that the company has not needed to add any new generating capacity in over 20 years in large part because its rate structures send the appropriate price signals to customers, who in turn have responded by conserving electricity during peak periods.

Distribution rates for commercial and industrial customers typically contain demand charges. The economic ef-

[4] The estimates conservatively assume that TOU or RTP rates have only twice the variability seen in average rates.

[5] Assuming that an ICE vehicle gets 30 mpg on \$3.50 per gallon gasoline and travels an average of 11,500 miles per year, the net savings per year for a PEV owner are \$1,169 if electric costs are \$0.05/kWh, \$997 if electric costs are \$0.10 /kWh, and only \$824 if electric costs are \$0.15/kWh at 300 Wh per mile.

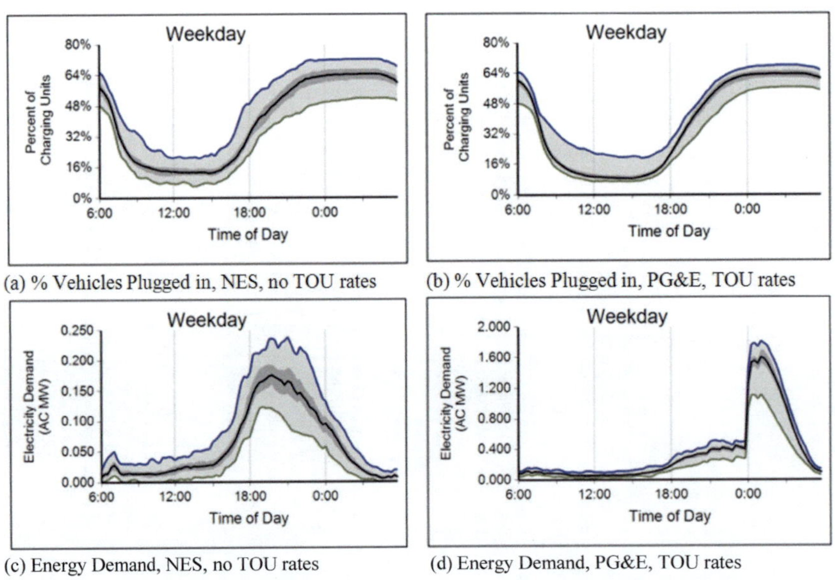

(a) % Vehicles Plugged in, NES, no TOU rates
(b) % Vehicles Plugged in, PG&E, TOU rates
(c) Energy Demand, NES, no TOU rates
(d) Energy Demand, PG&E, TOU rates

FIGURE 6-4 Residential charging behavior in NES and PG&E service territories, as measured in the EV Project. Panels (a) and (b) show average percent of vehicles plugged into residential chargers by time of day in the NES and PG&E service territories, and panels (c) and (d) show average charging energy demand by time of day in those territories. NOTE: NES, Nashville Electric Service; PG&E, Pacific Gas and Electric; TOU, time of use. SOURCE: ECOtality (2013).

fect on commercial or industrial customers that are providing charging could be substantial and strongly negative if a single hour with unusually high charging demand were to cause an increase in the demand charge. Although it might be argued that one or more charging stations would represent only marginal increases in energy consumption for relatively large commercial entities, to the extent that a charging station is being used during the peak power consumption time of day, it will have an impact on the maximum demand of the commercial or industrial entity. Exceeding the demand threshold by any amount will increase the total cost of energy to the facility and, in some cases, will hold the demand-rate charges at the higher level for many months to more than a year. A study done by the EV Project demonstrates the importance of this issue; it found that demand charges could account for over 90 percent of the utility bill in some areas (ECOtality 2012). Thus, it is critical to note that although the peak occurs only once and only for a brief period, the effect on the customer's bill could be felt for far longer, and more important, the increased cost could outweigh any potential benefits gained by providing PEV charging infrastructure.

There exists one additional impediment to PEVs that is directly related to the rate structure but difficult to quantify. PEVs individually and in combination with other technologies likely to be implemented in the distribution system (such as distributed storage, distributed generation, and advanced controls) might be able to provide a *benefit* to the utility in terms of ancillary services, such as regulation or reserves. The supply of those necessary services to the utility has a positive value in terms of cost savings—costs that the utility would have had to expend but for the fact that the PEV or other distributed device exists and is able to operate so as to benefit the utility. The ancillary service benefits are real, even if difficult to separate from the benefits of other technologies in the distribution system. The fact that PEVs and other technologies in the system can and do provide those services provides a positive benefit to the operations of the utility and could represent a financial benefit.[6] Although there is some difficulty in precisely quantifying the potential benefit, using the regulatory framework that exists in California would provide about \$100 per kW per year of capability and could be an important incentive for PEVs if passed on to PEV customers (E3 2014).[7] Regulatory structures implemented by PUCs and ISOs could allow these potential revenues to be claimed by PEVs. The existing operating and accounting logics implemented by PUCs and ISOs that allow customers to provide these services will need to be modified to accommodate PEVs, which are mobile loads that will be connecting at multiple and diverse locations as opposed to most (if not all) other distributed technologies that operate at a fixed location. Among those actions, the two most important are deciding which entity in the PEV ecosystem should be compensated for the service provided and how to measure compliance with any dispatch instruction given by the electric utility or ISO.

Finding: The confusion caused by the substantial differences in electric rates offered to customers by different utilities or states can be an impediment to PEV adoption.

Finding: TOU rate charging could provide a win-win situation as the PEV owner pays for charging at a lower rate and the utility benefits from moving the load from peak to off-peak.

Recommendation: To ensure that adopters of PEVs have incentives to charge vehicles at times when the cost of supplying energy is low, the federal government should propose that state regulatory commissions offer PEV owners the option of purchasing electricity under TOU or real-time pricing.

ELECTRICITY SECTOR REGULATORY ISSUES FOR OPERATING A PUBLIC CHARGING STATION

As noted in Chapter 5, utilities might have a viable business case for deploying charging infrastructure. Provision of PEV charging services can benefit electric utilities as it can increase utilization of fixed assets of the distribution infrastructure, potentially lowering rates, increasing revenues, or both. As noted, the provision of public charging might also encourage the adoption of PEVs, which could provide broad customer or societal benefits from reduced greenhouse gas emissions or improved local air quality. However, one caveat that needs to be recognized is that not all utilities are allowed to provide charging services. Some states have granted partial or full permission for electric utilities to provide PEV charging services. That action has allowed Austin Energy (Texas), Duke Energy (North Carolina), and Portland General Electric (Oregon) to fill the need for PEV charging services. In Japan, TEPCO is allowed to support the deployment of public charging infrastructure by providing necessary interconnection to the grid and internalizing the costs to the shareholders or ratepayers, and this approach has meaningfully reduced the cost to install DC fast chargers in TEPCO's service territory (Anegawa 2010).

Thus, many in the PEV and utility industries have called for greater latitude to provide charging services, particularly in underserved markets where demand for PEV charging exists (C2ES 2012). Independent public charging providers, however, have concerns about policies that would allow electric utilities to provide charging services and believe that the

[6] It has been suggested that benefits should be (and within most of the ISOs are) paid for based on the "avoided cost" of the utility. The difficulty is in calculating the avoided cost and therefore the size of any benefits. Avoided costs represent a calculation of what it would have cost the utility to acquire the service provided by the distributed technology if the utility had to provide it. A further difficulty in estimating this benefit is that the service is likely to be provided by multiple technologies within the distribution system, requiring sharing of any avoided-cost benefits, and that far more of the service could be delivered than the utility requires at any point in time.

[7] This estimate is the net present value over 10 years for a resource that would be available for the 100 peak hours of a year, assuming that the cost of a new entry has a weighted average cost of capital of an independent power producer.

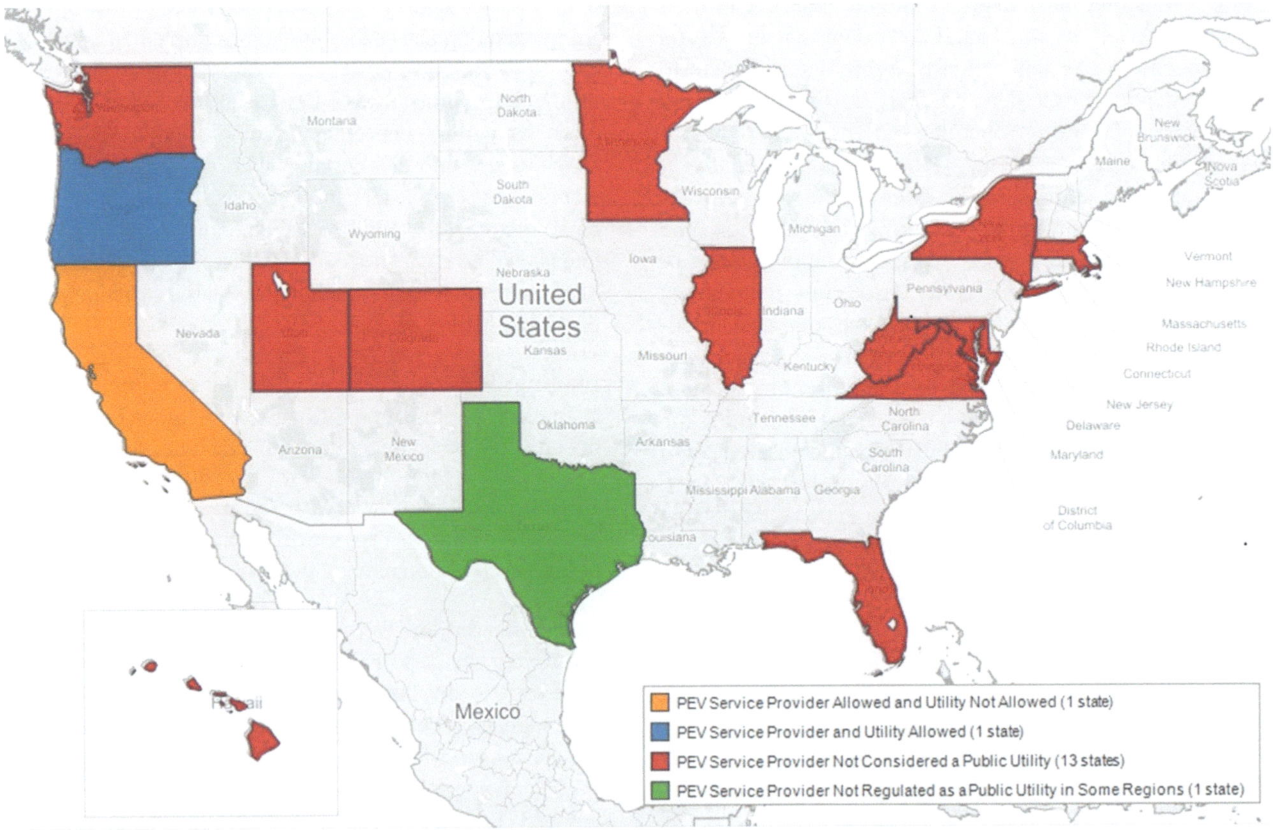

FIGURE 6-5 States that have regulations regarding who can own or operate a PEV charging station. NOTE: PEV, plug-in electric vehicle. SOURCE: Based on data from C2ES (2015). Courtesy of the Center for Climate and Energy Solutions.

utilities would have an unfair competitive advantage. Such policies could put independent public charging providers at a competitive disadvantage because utilities have substantial existing infrastructure and would be able to spread some of the cost of providing charging services across their customer base independent of whether any individual customer owned a PEV or used the public charging infrastructure.

Another regulatory issue is the extent to which PEV-charging providers are considered to be offering electricity for resale and thus would be regulated as a utility. As discussed above, in most states, the retail sale of electricity is a commercial activity heavily regulated as a monopoly business. Considering an independent public charging company to be a public utility and subject to public utility regulation would dramatically alter the company's cost structure and its potential competitive position. In addition, it would affect the company's ability to raise capital. Many states have not yet made a distinction between the retail sale of power and the provision of PEV charging services (Council of State Governments 2013). However, a few state PUCs have taken up the issue of whether PEV charging services should be a regulated activity,[8] and in a few states, the issue has been addressed by the legislature rather than by regulatory interpretation (see Figure 6-5).[9]

Finding: Electric utilities that provide PEV charging services have multiple reasons for doing so that can positively affect utility ratepayers and the utilities themselves.

Recommendation: As a means of encouraging consistency between jurisdictions, the federal government should propose that state regulatory commissions decide that public charging stations are not utilities and therefore not subject to utility regulatory oversight, specifically in setting rates for charging.

---

[8] For example, see California PUC Rulemaking 09-08-009, Code Section 740.2.14, July 2011; Arizona Corporation Commission Docket No. E-01345A-10-0123, Decision No. 72582, September 15, 2011; and PUC of Oregon Guidelines Adopted; Utilities Ordered To Make Revised Tariff Filings, January 19, 2012.

[9] For example, see Washington Substitute House Bill 1571, 62nd Legislature, 2011 Regular Session, July 22, 2011; California Assembly Bill 631; Colorado General Assembly House Bill 12-1258; and New York Bill S5110-2013.

Recommendation: Given that electric utilities and their ratepayers could benefit from increased PEV adoption, electric utility regulators should encourage their electric utilities to provide PEV charging services to their customers when conditions indicate that all customers benefit.

THE UTILITY OF THE FUTURE

The interactions of the electricity sector and PEV charging are not static or unidirectional. Internationally and now increasingly in the United States, the most significant changes in delivery of electricity are taking place on the customer's premises or inside the meter, where the customer has more control than the utility.[10] Such changes include programmable thermostats and smart appliances. There are also many changes occurring within the distribution system, including the introduction of micro-grids; the increased deployment of distributed electricity generation in the form of small-scale solar photovoltaics (PV) and wind; the development of distributed storage, including second-life PEV batteries; and advanced information technology and control.[11]

Consideration of PEVs in the future distribution system is an integral part of the ongoing planning that is focused on the utility of the future. PEV demand for charging energy will affect the total demand for energy at the distribution level. The increase in demand might well be offset by an increase in supply from distributed generation. Combining residential PV with the multiple possible functions of a PEV as a distributed storage device and means of transportation is also seen as a means of localized load balancing for the utility and cost savings for the customer.[12] In the future, increased information and communication technology combined with real-time economic price signals are anticipated to allow PEV battery systems to become distributed storage systems capable of providing energy and ancillary services to the distribution utility. Termed, variously, smart charging, vehicle-to-home, and vehicle-to-grid, this capability will give the distribution-system operator added flexibility and control to manage the overall load on the system.

Finding: PEVs might be a large part of the utility of the future and could help perform functions that the electric sector deems valuable. However, issues associated with customer access to their vehicles and effects on battery life will need to be resolved before vehicles can be fully integrated into the utility of the future.

---

[10] The customer's electric meter is where the "fence" is typically drawn, with the distribution company unable to see what happens on the customer premise beyond its meter.

[11] The MIT Energy Initiative study *The Utility of the Future* represents one research effort under way to understand the impact of disruptive technologies on the utility distribution system.

[12] It should be noted that PEV batteries beyond their useful life for transportation might be useable as stationary storage devices within the distribution system (see Chapter 4).

REFERENCES

Anegawa, T. 2010. Needs of Public Charging Infrastructure and Strategy of Deployment. TEPCO. Kanagawa, Japan.

CPUC (California Public Utilities Commission). 2012a. Load Research Report and Pricing Study. Alternative Fueled Vehicles Order Instituting Rulemaking R.09-08-009. San Francisco, CA. December 28.

CPUC. 2012b. "Workshop Presentation: Rate Design Elements, Concepts and Definitions." Residential Rate Structure Order Instituting Rulemaking R.12-06-013. December 5.

CPUC. 2014. Load Research Report. Alternative Fueled Vehicles Order Instituting Rulemaking R.09-08-009. San Francisco, CA. January 31.

Council of State Governments. 2013. State Utilities Law and Electric Vehicle Charging Stations. http://knowledgecenter.csg.org/kc/sites/default/files/Electric%20Vehicle%20Charging%20Stations_0.pdf.

C2ES (Center for Climate and Energy Solutions). 2012. An Action Plan to Integrate Plug-in Electric Vehicles with the U.S. Electrical Grid. Report of the PEV Dialogue Group. March. http://www.c2es.org/initiatives/pev/action-plan-report.

C2ES. 2015. Who Can Own/Operate a Charging Station? http://www.c2es.org/initiatives/pev/maps/who-can-own-operate-a-charging-station. Accessed March 24, 2015.

DOT (U.S. Department of Transportation). 2014. "National Transportation Statistics." http://www.rita.dot.gov/bts/sites/rita.dot.gov.bts/files/publications/national_transportation_statistics/index.html.

E3 (Energy and Environmental Economics). 2014. California Transportation Electrification Assessment Phase 2: Grid Impacts, October 23. http://www.caletc.com/wp-content/uploads/2014/10/CalETC_TEA_Phase_2_Final_10-23-14.pdf.

ECOtality. 2012. Lessons Learned—The EV Project White Paper Syllabus. Prepared for the U.S. Department of Energy Award #DE-EE0002194. http://www.theevproject.com/downloads/documents/Syllabus%20Rev%204.0.pdf.

ECOtality. 2013. "How Do PEV Owners Respond to Time-of-Use Rates while Charging Vehicles?" http://www.theevproject.com/cms-assets/documents/125348-714937.pev-driver.pdf.

EEI (Edison Electric Institute). 2014. Transportation Electrification: Utility Fleets Leading the Charge. Washington, DC. June.

EIA (U.S. Energy Information Administration). 2011. "About 60% of the U.S. Electric Power Supply Is Managed by RTOs." Today in Energy, April 4. http://www.eia.gov/todayinenergy/detail.cfm?id=790.

EIA. 2012. "Table A1. Total Energy Supply and Disposition Demand" and "Table A2. Energy Consumption by Sector and Source." Annual Energy Outlook 2013 Early Release Overview. December 5. http://www.eia.gov/forecasts/aeo/er/index.cfm. Accessed March 14, 2013.

EIA. 2013. Annual Energy Outlook 2013 Market Trends—Electricity, May. Figure 75. http://www.eia.gov/forecasts/aeo/pdf/0383%282013%29.pdf.

EIA. 2014. Electric Power Monthly with Data for December. February. http://www.eia.gov/electricity/monthly/current_year/february2014.pdf.

FERC (Federal Energy Regulatory Commission). 1996. Promoting Wholesale Competition Through Open Access Non-discriminatory Transmission Services by Public Utilities; Recovery of Stranded Costs by Public Utilities and Transmitting Utilities. Order No. 888. Issued April 24.

INL (Idaho National Laboratory). 2014. EV Project Electric Vehicle Charging Infrastructure Summary Report: October 2013 through December 2013, INL/MIS-10-19479. http://avt.inel.gov/pdf/EVProj/EVProjectInfrastructureQ42013.pdf.

Kind, P. 2013. Disruptive Challenges: Financial Implications and Strategic Responses to a Changing Retail Electric Business. Edison Electric Institute. January.

Maitra, A. 2011 Potential Impacts of Vehicles on the Grid. Electric Power Research Institute. Palo Alto, CA. August.

Maitra, A., J. Taylor, D. Brooks, M. Alexander, and M. Duvall. 2009. Integrating Plug-in Electric Vehicles with the Distribution System. 20th International Conference on Electricity Distribution, Prague, June 8-11. Electric Power Research Institute. Palo Alto, CA.

U.S.-Canada Power System Outage Taskforce. 2004. Final Report on the August 14, 2003 Blackout in the United States and Canada: Causes and Recommendations. April. http://energy.gov/sites/prod/files/oeprod/Documentsand-Media/BlackoutFinal-Web.pdf. Accessed April 23, 2013.

U.S. Court of Appeals. 2014. Electric Power Supply Association, Petitioner v. Federal Energy Regulatory Commission, Respondent and Madison Gas and Electric Company et al., Intervenors. No. 11-1486. May 23. http://www.cadc.uscourts.gov/internet/opinions.nsf/DE531DBFA7DE1ABE85257CE1004F4C53/$file/11-1486-1494281.pdf.

7

Incentives for the Deployment of Plug-in Electric Vehicles

One of the most important issues concerning the deployment of plug-in electric vehicles (PEVs) is determining what, if any, incentives are needed to promote this deployment. Determining appropriate incentives is difficult because little is yet known about the effectiveness of PEV incentive programs. Therefore, the committee first considered the price or cost competiveness of PEVs and the possibilities for reducing production costs. It next considered manufacturer and consumer incentives for purchasing or owning PEVs and then past incentive programs for other alternative-vehicle and fuel technologies. The chapter concludes with recommendations on what the committee sees as the most compelling approaches to promoting PEV deployment.

VEHICLE PRICE AND COST OF OWNERSHIP

A major consideration when purchasing or leasing a vehicle is the financial consequence. Thus, the two factors to consider are the vehicle price and the cost of ownership. Most people compare prices or monthly leasing payments and take into account any financial incentive that would reduce the vehicle price. Because vehicle manufacturers and dealers are profit-oriented businesses, vehicle prices are generally related to *production costs*. However, the relationship between a manufacturer's suggested retail price and its production cost typically reflects a number of considerations and might differ across vehicles and over time. For example, on newly developed vehicles, such as PEVs, vehicle manufacturers might be motivated to incur losses or relatively low profit margins in the short run to promote sales and strengthen their business positions and profit margins over the long run. That type of marketing strategy contributed to the eventual success of the Toyota Prius (Tellis 2013). Regardless of how the vehicle price is set, price is an important consideration for most consumers when shopping for a new vehicle.

Some prospective buyers also consider more broadly the costs of owning a vehicle—in particular, the costs of fueling, maintaining, and insuring the vehicle and its resale or trade-in value. The *total cost of ownership* (or overall cost to the consumer) of any specific vehicle can be viewed as the effective purchase price (price adjusted for any financial incentives) plus the costs of fueling, maintaining, and insuring the vehicle, minus the resale or trade-in value.[1] Although the distinction between price and total cost of ownership is important, most consumers find it difficult to estimate the latter with much confidence, partly because they are not certain at the time of purchase how long they will keep their vehicles or how many miles they will drive in them and partly because fuel costs, maintenance costs, and resale values are uncertain, particularly for newer technology vehicles like PEVs. That said, tools to help prospective buyers estimate and compare the total ownership costs of different vehicles are now available at Edmunds.com and elsewhere.

The U.S. Department of Energy (DOE) has created a website calculator where prospective buyers provide various driver-specific inputs (such as net vehicle price, normal daily driving distance, annual mileage, and breakdown of mileage between city and highway) and can calculate the *cumulative cost of ownership* over different time horizons under representative assumptions about other such factors as maintenance and insurance costs (DOE 2014a).[2] Cumulative cost of ownership is distinct from total cost of ownership in that it is a calculation for a given time horizon and typically does not include the trade-in or resale value. Thus, by focusing on the cumulative costs of ownership over different time horizons, the DOE calculations avoid assumptions about resale values, which are highly uncertain for PEVs.

Affordability—as reflected in price and total cost of ownership—is not the only consideration that influences the types of vehicles that consumers choose. As emphasized in Chapter 3, PEV adoption also depends importantly on consumer awareness, the variety of models available, uncertainties about new technologies and resale values, and various vehicle attributes that determine its utility to the customer. Because there is still much uncertainty about PEV technolo-

[1] More sophisticated definitions of the total cost of ownership are based on the present discounted values of the various components of cost (that is, they discount future costs relative to current costs).

[2] The website calculator provides zip-code-specific assumptions about fuel prices that the user can override.

gies and the battery lifetimes, and because there are not as yet well-developed markets for used PEVs or their batteries, consumers are likely to perceive more uncertainty about the total costs of owning PEVs than about the total costs of owning conventional vehicles. Those uncertainties make risk-averse consumers less likely to purchase a PEV, other things being equal. They also strengthen the incentive to lease a PEV rather than purchase one.

Thus, leasing is a more frequent choice for PEVs than for conventional vehicles (see Table 3-3) because it can make monthly payments for the vehicle appear more affordable and reduce the risk of owning one. In a typical leasing arrangement, ownership of the vehicle is transferred at a negotiated sales price from the dealer to a bank or some other finance company (often the financial arm of the vehicle manufacturer). In general, the sales price is influenced importantly by the amount that the vehicle manufacturer charges the dealer for the car and by any incentive payments or guarantees that the manufacturer offers the finance company.[3] The finance company collects the monthly leasing payments and generally also receives (1) incentives or other subsidies provided by federal or state governments or (2) benefits from lower negotiated sales prices resulting from government incentives provided to manufacturers or dealers. As such, potential PEV drivers who do not have enough income to qualify for the federal income tax credit could still benefit from any credit available with vehicle leasing.

PRICE AND COST COMPETITIVENESS OF PLUG-IN ELECTRIC VEHICLES

The committee used three approaches to assess the extent to which prices or costs of ownership might have affected PEV deployment to date. First, the committee compared the manufacturers' suggested retail prices (MSRPs, which are essentially the target prices) of various PEV models with those of comparative vehicles. Second, it evaluated sales data, and, third, it considered consumer surveys. No approach provided conclusive results.

Table 7-1 lists MSRPs for three relatively best-selling PEV models, for several hybrid electric vehicles (HEVs), and for internal-combustion engine (ICE) vehicles. It excludes the Tesla Model S, which tends to be bought by relatively wealthy individuals whose purchase decisions might not be highly sensitive to price. The table allows one to compare the prices of the Chevrolet Volt and two Chevrolet Cruze models and to compare the prices of the Ford Fusion Energi and the Ford Fusion Automatic and Hybrid models; the comparisons are informative because the PEVs and comparative ICE and HEV models are built on the same platforms and therefore have similar production costs for components other than those associated with their sources of energy and drive trains.[4] The table also includes the prices of the Toyota Prius, the Volkswagen Passat, and the Nissan Leaf, along with the average transaction prices for small and midsize vehicle segments, including the prices for the specialty segment. The committee emphasizes that the MSRPs shown in the table simply represent price points that manufacturers target, as distinct from the prices at which vehicles are actually sold, which are typically less than the MSRPs. Nevertheless, the MSRP comparisons provide some useful perspectives. The average transaction price data reflect the actual prices paid by the consumer.

In addition to providing information on vehicle prices, Table 7-1 includes ranges for annual fuel costs and 5-year cumulative costs of ownership. As reflected in the table notes, the estimates are based on a combination of the assumptions included in the DOE calculator and the committee's specific assumptions about annual vehicle miles traveled, electricity costs, and gasoline prices ranging from \$2.50 to \$4.00 per gallon. The estimates of 5-year cumulative costs of ownership assume that purchasers pay MSRPs and receive maximum tax credits, and the estimated costs in years two through five have not been discounted.[5]

As indicated in Table 7-1, the MSRPs before consideration of the federal tax credits for the PEVs are all substantially higher than the MSRPs for the HEVs and ICE vehicles listed in the table. After consideration of the \$7,500 federal tax credit, the adjusted MSRP for the Chevrolet Volt still substantially exceeds the MSRPs for the Chevrolet Cruze LS Automatic, the Toyota Prius, and the Volkswagen Passat and the average transaction price for the specialty small vehicle segment. It somewhat exceeds the MSRP of the Chevrolet Cruze Diesel Automatic. The MSRP for the Ford Fusion Energi, after adjusting for the \$4,007 federal tax credit, exceeds MSRPs of the two other Fusion models, the Toyota Prius, and the Volkswagen Passat but is similar to the average transaction price of the specialty midsize vehicle segment. The 5-year cumulative cost of owning a Chevrolet Volt—as estimated by the DOE calculator using representative assumptions—is, respectively, about \$2,300 and \$3,400 higher than the comparable 5-year costs for the Toyota Prius and the Volkswagen Passat at a gasoline price of \$2.50 per gallon and higher by

[3] An important influence on the negotiated sales price is the finance company's estimate of the value of the car at the end of the lease, taking into account any guarantees by the vehicle manufacturer.

[4] It is common practice for a vehicle manufacturer to build multiple vehicle models on the same platform but that are in different market segments. Although the Chevrolet Cruze and Volt are in the same size segment, they are not in the same market segment (standard compact vs premium compact).

[5] In theory, car payments that are spread over 5 years or less have present discounted values that equal, or closely approximate, the vehicle MSRP when the payments are discounted at the rate of interest charged in financing the car payments. And while discount rates between 0 and 6 percent would imply that the present discounted value of the 5-year stream of other costs (for fuel, tires, maintenance, insurance, inspection, and registration) was up to several thousand dollars less than five times the annual average of those costs—with relatively smaller differences for vehicles that have relatively lower annual fuel costs (such as PEVs)—the qualitative comparisons and the finding would not be affected.

TABLE 7-1 MSRPs and 5-Year Cumulative Cost of Ownership for Selected Plug-in Electric Vehicles and Comparative Vehicles (dollars)

Model[a]	Vehicle Type	MSRP[b]	MSRP Less Federal Tax Credit[c]	Range of Annual Fuel Cost[d]	Range for 5-Year Cumulative Cost of Ownership[d,e]
Chevrolet Volt	PHEV	34,185	26,685	540-636	40,564-41,038
Chevrolet Cruze LS Automatic (6S)	ICE vehicle	19,530	19,530	915-1,464	35,280-38,025
Chevrolet Cruze Diesel Automatic (6S)	ICE vehicle	25,810	25,810	990-1,485	41,935-44,410
Ford Fusion Energi	PHEV	34,700	30,693	657-897	45,153-46,353
Ford Fusion Automatic	ICE vehicle	22,400	22,400	1,002-1,605	38,589-41,598
Ford Fusion Hybrid (FWD)	HEV	27,280	27,280	680-1,088	41,855-43,895
Nissan Leaf	BEV	29,010	21,510	435	34,860
Toyota Prius Hybrid	HEV	24,200	24,200	581-932	38,284-40,033
Volkswagen Passat (6S)	ICE vehicle	20,995	20,995	1,005-1,608	37,195-40,210
Alternative Comparators	Average Transactions Price[f]				
Small vehicles (average)	20,374				
Small vehicles (specialty)	23,129				
Midsize vehicles (average)	25,677				
Midsize vehicles (specialty)	29,759				

[a] All HEV and ICE vehicle models are ones with four-cylinder engines.
[b] Based on MSRP data for 2014 models from manufacturers' websites as of July 31, 2014, except for Volkswagen Passat, which is as of August 31, 2014.
[c] The federal tax credit is $2,500 for PEVs that have battery capacities below 5 kWh. For PEVs that have larger battery capacities, the credit is set at $2,500 plus $417 times the amount that the battery capacity exceeds 5 kWh, up to a maximum of $7,500. The subsidy extends to the first 200,000 PEVs sold by each manufacturer. Those purchasers who would not otherwise pay at least $7,500 in federal income taxes cannot take advantage of the full $7,500 credit. See Internal Revenue Code Section 30D (IRS 2009).
[d] Based on 11,500 vehicle miles traveled, $2.50-$4.00 per gallon for gasoline, $3.00-$4.50 per gallon for diesel fuel, $0.125 per kWh for electricity, and DOE assumptions about normal daily use.
[e] Includes DOE estimate of $2,235 annual costs for tires, maintenance, insurance, licensing, and registration when vehicle is driven 11,500 miles. Excludes any cost of financing and any costs associated with home-charger installation and permitting requirements. Assumes purchaser pays MSRP and receives maximum tax credit.
[f] Data provided by Baum & Associates (A. Baum, personal communication, April 22, 2014).
NOTE: BEV, battery electric vehicle; FWD, front-wheel drive; HEV, hybrid electric vehicle; ICE, internal-combustion engine; MSRP, manufacturers' suggested retail price; PHEV, plug-in hybrid electric vehicle; S, speed.

about $1,000 and $800 at a gasoline price of $4.00 per gallon. The 5-year cumulative cost of owning a Ford Fusion Energi is higher than those of all the HEV and ICE vehicles listed in Table 7-1, even at a gasoline price of $4.00 per gallon.

In contrast, the MSRP of the Nissan Leaf, after considering the $7,500 federal tax credit, is less than the MSRP of the Toyota Prius and is comparable to the average transaction price of the small vehicle segment. The 5-year cumulative ownership cost of the Nissan Leaf, as estimated using the DOE calculator, is, respectively, about $3,400 and $2,300 less than the analogous costs of the Toyota Prius and the Volkswagen Passat at a gasoline price of $2.50 per gallon and about $5,200 and $5,400 less at a gasoline price of $4.00 per gallon.

A second approach to assessing the price-competitiveness of PEVs is to evaluate data on sales volumes. The idea is that low sales volume could indicate that the vehicles are not price- or cost-competitive, although the analysis above indicates that the Nissan Leaf is highly competitive given the federal tax credit. As noted in Chapter 1, about 290,000 highway-capable PEVs were sold in the United States by the close of 2014. Although there is no generally accepted wisdom on how rapidly sales of a new product line should expand during the early introductory stage, the number of PEVs sold in the United States to date has fallen short of aspirational goals,[6] despite substantial incentives.

Consumer survey data provide yet another approach for evaluating whether price is currently an obstacle to PEV deployment. The annual New Vehicle Experience Studies conducted by Strategic Vision have surveyed large samples of new-vehicle buyers and distinguished between those who actively looked at PEVs but chose not to purchase one, and those who did not even consider purchasing a PEV. In the former group, 41 percent indicated that current prices or rebates were appealing and 27 percent said that current interest or lease rates were appealing; in the latter group, 25 percent considered them appealing and 17 percent said that current interest or lease rates were appealing (Edwards 2013). The survey data also showed that 43 percent and 45 percent of BEV and PHEV buyers, respectively, considered their purchases "value for the money." The data suggest that current pricing might not be a primary barrier to greater PEV sales among early adopters, who tend to be less price sensitive than the mainstream market, and that buyers are rejecting the vehicles for other reasons. Strategic Vision also found that 10 percent of new-vehicle buyers are actively shopping or plan to shop for a PEV and that about 33 percent are open to hearing more about what PEVs can do for them. Given that 15 million new vehicles are sold each year, the data suggests an active potential market of about 1.5 million PEVs with about two-thirds finding the pricing or lease rates appealing. Despite these results, Strategic Vision stated that it still believes that "price is a critical barrier—even for more affluent customers" (Edwards 2013, p. 45).

[6] For example, in early 2011 DOE projected cumulative U.S. sales of 1.22 million PEVs by 2015 (DOE 2011).

Finding: Under the current program of federal tax credits, the comparisons of MSRPs and cumulative ownership costs provide mixed evidence on whether price is currently an obstacle to the deployment of PEVs. However, in the absence of the tax credits or other subsidies, analogous comparisons at prevailing MSRPs would be unfavorable to the PEVs.

Finding: Sales data and consumer survey data are difficult to interpret. They are consistent, however, with the view that price is a barrier to some buyers, but that others might be rejecting PEVs for other reasons.

POSSIBILITIES FOR DECLINES IN PRODUCTION COSTS FOR PLUG-IN ELECTRIC VEHICLES

The extent to which PEVs are adopted over time will depend on reductions in their production costs, on the policies that governments implement to promote PEV deployment, and on the extent to which vehicle manufacturers decide to price PEVs more attractively by relying on relatively low markups in pricing and perhaps compensating for the loss of revenue by raising mark-ups on their portfolios of other vehicles.[7] The three factors are not completely independent. Government policies toward research and development can affect battery costs, and policies, such as zero-emission requirements, can induce vehicle manufacturers to change their pricing strategies. This section focuses on likely reductions over time in the production costs of PEVs.

In general, the costs of producing PEVs will be driven down over time by a number of factors. The technologies being used for PEVs are relatively new compared with technologies used to produce ICE vehicles, which have been evolving and improving for more than a century. Thus, as discussed in Chapter 2, it is expected that research and development will lead to reductions in the costs of PEV batteries over time through technological improvements, such as higher energy densities, improved designs, and longer battery lives.[8]

[7] Manufacturers typically estimate their direct labor and material costs of production and markup prices above direct production costs by amounts sufficient to cover the fixed costs of plant and equipment, various indirect costs (including the costs of research and development, corporate operations, dealer support, and marketing), and an allowance for profits. A study based on data from the 2007 annual reports of eight major vehicle manufacturers found that the average markup factor for the automobile industry was about 1.5 (RTI/UMTRI 2009). However, manufacturers will sell vehicles at prices that the market will bear; thus, the markup on one vehicle model can be much greater than that on another vehicle model.

[8] Innovations in other elements of vehicle technology are likely to lead to improved vehicle performance without necessarily generating substantial reductions in cost. Improving the aerodynamics, reducing friction, reducing the rolling resistance of tires, and reducing weight could lead to a vehicle design with, for example, better performance or more driving range, depending on what trade-offs were made in the overall vehicle design. That could lead to the need for a smaller battery, and the reduced cost of the battery would have to be weighed against the increased cost of the above-mentioned improvements because such improvements generally come at

Cost reduction might also be realized as PEV production volume increases and both the capital costs of investments in production facilities and the indirect costs of research and development, corporate operations, dealer support, and marketing are spread over more vehicles. Such scale economies will also be realized in the supplier base with increases in the demand for components, such as batteries, motors, and inverters, enabling suppliers to reduce the prices that vehicle manufacturers are charged for those components. In addition, increases in the number of vehicles and the demand for components will likely lead to greater competition among suppliers, which could intensify the downward pressure on component prices as suppliers innovate and generate better designs. And once a new vehicle technology becomes fairly firmly established, components tend to become more standardized, leading to additional reductions in production costs.

As emphasized in Chapter 2, the difference between the costs of producing PEVs and comparative conventional vehicles can be largely attributed to the high cost of high-energy batteries. Accordingly, the prospect for large-scale deployment of PEVs depends importantly on how much battery costs decline over time. Other things equal, if battery pack costs declined by as much as 50 percent over the next 5 to 10 years, consistent with optimistic projections (see discussion in Chapter 2), the cost of producing a BEV with 24 kWh nominal battery capacity (analogous to the Nissan Leaf) would decline by roughly $6,000. Similarly, the costs of producing PHEVs with 16.5 kWh and 7.6 kWh nominal battery capacities (analogous to the Chevrolet Volt and Ford Energi) would decline by about $4,100 and $1,900. And a 75 percent decline in battery pack costs (a highly optimistic forecast)—to as low as $125 per kWh of nominal battery capacity—would reduce the costs of producing the Nissan Leaf, the Chevrolet Volt, and the Ford Energi by an additional $3,000, $2,050, and $950, respectively. Such optimistic reductions in production costs would provide opportunities for the vehicle manufacturers to reduce the MSRPs for PEVs by amounts that largely offset, or more than offset, the effects of the pending expiration of the current program of federal tax credits.

A detailed analysis of how the nonbattery costs of PEVs are likely to evolve relative to those of comparative vehicles is beyond the scope of this report. Recent NRC reports on the costs of different vehicle types, however, conclude that the costs of producing PHEVs and BEVs will likely remain greater than the costs of producing ICE vehicles and HEVs for at least the next two decades (NAS/NAE/NRC 2009; NRC 2013b).

It should be noted that PEV adoption does not require PEVs to be priced at or below the prices of conventional vehicles. Some might buy PEVs because they are less expensive to fuel and perhaps less expensive to maintain. Others might value certain of their attributes that are not present (or not present to the same degree) on conventional vehicles, such as the smooth and quiet ride, better acceleration, the convenience of home fueling, less maintenance (fewer or no oil changes), and the potential for reducing vehicle emissions and petroleum usage.

---

some incremental cost (NRC 2011a, 2013a). These innovations in technology will likely also be applied to conventional vehicles as manufacturers strive to meet fuel-economy requirements, but also at some incremental cost.

Finding: Although battery costs could decline by 50 or perhaps even 75 percent over the next decade, it is not clear whether such a decline would be sufficient—by itself—to ensure widespread adoption of PEVs once the current quotas for federal tax credits are exhausted.

Finding: The decline over time in PEV production costs is likely to occur gradually, and existing quotas for federal tax credits might be exhausted for manufacturers of relatively popular PEVs before costs can be substantially reduced.

INCENTIVES

The production and purchase of PEVs is a classic chicken-and-egg problem. Manufacturers do not want to produce PEVs if no customers exist, and consumers cannot buy PEVs if vehicles are not available that meet their expectations. Therefore, regulatory requirements and incentives for manufacturers and consumers have been provided over the past few years by states and the federal government to encourage PEV production and deployment. Most manufacturer incentives and mandates are contained in federal or state regulatory programs discussed below. Most consumer incentive programs described below have involved purchase incentives, although some have included ownership and use incentives. There have also been incentives to install charging stations, the availability of which might also influence people's willingness to purchase PEVs.

Manufacturer Incentives and Regulatory Requirements

Incentives for manufacturers to produce PEVs are contained in the federal Corporate Average Fuel Economy (CAFE) Standards and the Greenhouse Gas (GHG) Emission Standards for light-duty vehicles. California and other states have Zero-Emission-Vehicle (ZEV) programs that require the sale of PEVs in those states because PEVs are the only qualifying technology that are currently mass produced. These manufacturer incentives and regulatory requirements can have the effect of reducing the vehicle price of PEVs relative to other vehicles; they are reviewed in detail below.

Federal Regulatory Incentives for Plug-in Electric Vehicles

Fuel economy and GHG emissions from light-duty vehicles are regulated under the federal CAFE-GHG national program. Under the recently updated rule, vehicle manufacturers must comply with fuel economy and GHG standards that are equivalent to about 54.5 mpg and 163 grams of carbon dioxide (CO_2) per mile for the fleet average of new vehicles by model year (MY) 2025 (EPA/NHTSA 2012a). Although GHG emissions from light-duty vehicles are regulated by the U.S. Environmental Protection Agency (EPA) under the Clean Air Act and fuel economy is regulated by the National Highway Traffic and Safety Administration (NHTSA) under the CAFE program, the federal agencies have developed a single national program.[9] The standards are fleet-based standards, meaning that a manufacturer can build vehicles that are certified above and below the standards as long as the fleet-wide average meets the standards. The standards also offer an array of regulatory flexibilities, including the ability to bank or buy compliance credits and incentives for various types of technologies. Although the analyses done by EPA and NHTSA for their most recent regulation for 2017-2025 developed a cost-effective compliance demonstration pathway that shows how the standards for vehicles in 2025 can be achieved almost exclusively through conventional gasoline-powered vehicles, the regulations developed by EPA and NHTSA have generous credits for PEVs that make it attractive for vehicle manufacturers to produce PEVs. However, the very nature of the separate legislative authorities under which EPA and NHTSA operate to regulate light-duty vehicles means that the manner of crediting manufacturers of alternative-fuel vehicles, such as PEVs, diverges between the CAFE and GHG standards.

The CAFE standard focuses on reducing petroleum usage in the United States. Federal law requires the CAFE program to evaluate PEVs and all other alternative-fuel vehicles by using a petroleum equivalency factor (PEF) to calculate a fuel economy compliance number for such vehicles.[10] PEFs are used to convert the electric energy consumption measured in the certification test cycle of alternative-fuel vehicles, including BEVs and PHEVs, to an equivalent fuel economy number. Manufacturers use the miles per gallon equivalent (MPGe) to calculate their fleet average mpg for compliance purposes. In compliance with the law, only 15 percent of the alternative fuel (such as electricity) that is consumed during the test is counted toward the fuel economy rating of an alternative-fuel vehicle. That treatment provides a strong incentive for manufacturers to produce alternative-fuel vehicles to comply with CAFE program requirements. For example, a BEV that is rated on the certification test cycle at 230 Wh/mile (roughly equivalent to the certification test cycle value for a Nissan Leaf) is treated as equivalent to a 357 mpg gasoline-powered car (see Box 7-1).

PHEVs are treated as dual-fuel vehicles that use both electricity and gasoline. Federal regulations stipulate how the gasoline and electric energy consumption is measured in certification test cycles for PHEVs. The measured electric energy consumption is converted, as in the BEV case, to an MPGe by using the petroleum equivalency factor method. The electric MPGe and gasoline mpg must be weighted to obtain the fuel economy value used in CAFE compliance. Until 2019, PHEVs are assumed to use electric fuel 50 percent of the time and gasoline 50 percent of the time (EPA/NHTSA 2012b). Beginning in MY 2020, the weighting will be determined on the basis of the SAE J1711 fuel economy test method that uses a utility factor to estimate the fraction of driving on electricity and assumes that the vehicle owner charges once per day and drives in much the same way as today's typical light-duty vehicle drivers. Given that method, a PHEV with 20-mile all-electric range (PHEV20) would be treated as a 90 mpg gasoline-powered car, and a PHEV with 60-mile all-electric range (PHEV60) would be treated as a 226 mpg gasoline-powered car (Al-Alawi and Bradley 2014).

The EPA GHG standards provide two temporary incentives to vehicle manufacturers to produce PEVs. The first incentive is temporary treatment of PEVs as zero emissions (that is, upstream emissions of power plants are ignored) for the portion of operation assumed to be powered by electricity. For BEVs and fuel-cell vehicles (FCVs), the MY 2017-2021 GHG standards set a value of 0 g/mile for the tailpipe CO_2 emissions compliance value (EPA/NHTSA 2012b). PHEVs also receive a value of 0 g/mile based on a formula to estimate the fraction of electricity usage. For MY 2022-2025, the program allows the 0 g/mile treatment up to a cumulative sales cap for each manufacturer.[11] After that cap is reached, the compliance values for BEVs and the electric portion of PHEVs are based on an estimate of the national average emissions associated with producing the electricity needed to charge PEVs. However, the cumulative sales caps appear to be generous, so it is possible that most PEVs will be treated as ZEVs.

The second manufacturer incentive under the EPA GHG standards is sales multipliers that effectively treat a single PEV sold as more than one vehicle for compliance purposes. The PEV sales multipliers start at 2.0 in MY 2017 for BEVs and FCVs and 1.6 for PHEVs and then gradually decline to 1.0 by MY 2022, when they are proposed to be completely phased out. The larger multiplier in the earlier years rewards manufacturers that are early market leaders. Allowing each PEV to count as more than one vehicle lowers the average GHG per mile for a manufacturer.

By increasing the MPGe and decreasing the grams CO_2 per mile of PEVs, the federal incentives from the PEF, zero

[9] The largest source of GHG emissions from light-duty vehicles is CO_2 that results from the combustion of gasoline or diesel fuel, and this implies that fuel economy and GHG emissions are directly correlated, necessitating the development of a common set of standards.

[10] 49 U.S.C. 32904(a)(2)(B) and 49 U.S.C. 32905(a).

[11] Manufacturers that sell 300,000 PHEVs, BEVs, and FCVs between 2019 and 2021 can use the 0 g/mile value for a maximum of 600,000 vehicles starting in 2022. For all other manufacturers (those who sell less than the 300,000), the 0 g/mile value can be used only up to 200,000 vehicles. After the sales cap is reached, emissions will be calculated using an upstream emission standard calculated by EPA.

BOX 7-1 Derivation of Petroleum Equivalent for a Battery Electric Vehicle

The PEF is derived by first calculating a full fuel cycle, gasoline-equivalent energy content of electricity and then dividing it by a 0.15 fuel-content factor. The PEF was developed to motivate the production of vehicles fueled with 85 percent ethanol (E85), and the 0.15 factor reflects the petroleum consumption of E85 vehicles. The gasoline-equivalent energy content of electricity (E_g) is calculated as follows:

$$E_g = \text{gasoline-equivalent energy content of electricity} = (T_g \times T_t \times C) / T_p$$

where

T_g = U.S. average fossil-fuel electricity generation efficiency = 0.328

T_t = U.S. average electricity transmission and distribution efficiency = 0.924

T_p = petroleum refining and distribution efficiency = 0.830

C = watt-hours of energy per gallon of gasoline conversion factor = 33,705 Wh/gal

Therefore,

$$E_g = (0.328 \times 0.924 \times 33{,}705)/0.830 = 12{,}307 \text{ Wh/gal}$$

The Nissan Leaf, which requires 230 Wh/mile, exhibits a range of 53.5 miles on the electric-energy equivalent of 1 gallon of gasoline (12,307/230 = 53.5). That is the certification test-cycle result. To provide an incentive for alternative-fuel vehicles, only 15 percent of the fuel consumed in the test cycle is counted, and the resulting MPGe for the Nissan Leaf is 12,307/(230 × 0.15) = 357 mpg for CAFE purposes.

emissions treatment, and sales multipliers allow the manufacturers to produce higher emitting and less fuel-efficient gasoline-vehicle fleets and still meet their fleet average standards. The incentives, therefore, create an internal cross subsidy that allows a manufacturer to reduce the cost of compliance for their gasoline-vehicle fleet by producing PEVs. Furthermore, because credits can be traded between manufacturers, such companies as Tesla that produce excess CAFE and GHG credits can sell their credits to other manufacturers (Energy Independence and Security Act 2007). The value of the PEV credits under EPA and NHTSA regulations is difficult to estimate but might be about a few thousand dollars per vehicle based on the costs of regulatory compliance in the absence of PEVs (EPA/NHTSA 2012b). The California and state ZEV programs (see below) also generate credits for various attributes, and the value has been estimated at up to $35,000 per vehicle for the Tesla Model S, which generates up to seven credits per vehicle sold (Ohnsman 2013).

State Zero-Emission-Vehicle Programs

The California ZEV program provides an important manufacturer requirement for PEVs. The Clean Air Act Amendments of 1969 authorized California to develop more stringent tailpipe standards than the rest of the country, and the California Air Resources Board (CARB) used the authority provided in Section 209b to adopt the original ZEV program in 1990. The ZEV program is a part of the state's comprehensive plan to meet federal and state ambient air quality standards. Later amendments to the Clean Air Act allowed other states to opt into the California standard. Nine states—Connecticut, Maine, Maryland, Massachusetts, New Jersey, New York, Rhode Island, Vermont, and Oregon—have adopted the California ZEV program as authorized under Section 177 of the Clean Air Act as part of their plans to meet federal ambient air quality standards. The nine states and California account for 28 percent of total U.S. light-duty vehicle sales. Recently, eight states that have the ZEV program signed a joint memorandum of agreement to cooperate on developing policies to accelerate PEV deployment in their states (State ZEV Programs 2013). The agreement includes the development of a Multistate ZEV Action Plan that describes state actions to promote PEV deployment and recommends research and stakeholder partnerships to support long-term development of the PEV market (ZEV Program Implementation Task Force 2014). Additionally, many of the ZEV states participate in the Northeast Electric Vehicle Network, which works to promote PEV deployment.

For a manufacturer to receive credit for vehicle sales under the ZEV program, a vehicle must be categorized under the program as either a ZEV (a BEV or an FCV) as defined by the program or a PHEV with an all-electric range of greater than or equal to 10 miles (CARB 2013). CARB estimates that the total number of ZEVs (BEVs and FCVs) and PHEVs needed to comply for MYs 2018 through 2025 for California and the nine other states is about 228,000 in 2018 and 725,000 by 2025 (Keddie 2013). The ZEV Program Implementation Task Force (2014) predicts that by 2025, a little more than 15 percent of new vehicles sold in participating states will be either ZEVs or PHEVs.

Finding: By creating an internal cross subsidy, existing federal and state regulatory programs for fuel economy and emissions (CAFE, GHG, and ZEV) have been effective at stimulating manufacturers to produce some PEVs. The sale of credits from these programs between manufacturers has also provided an important incentive for PEV manufacturers to price PEVs more attractively.

Finding: Because the ZEV program mandates sales of a certain percentage of PEVs, its impact could be larger than the incentives under the federal CAFE-GHG national program.

Consumer Incentives

The U.S. federal, state, and local governments have all experimented with consumer incentives to encourage PEV deployment. Many other countries have also used various policy tools to encourage consumer adoption. Box 7-2 defines the different types of financial incentives that have been used, and Table 7-2 summarizes the various financial and nonfinancial incentives and the entities that have used them. The incentives that have been used to promote PEV deployment are discussed below.

The committee defined four categories of financial incentives. *Purchase incentives* are one-time financial benefits earned by purchase of a PEV and include tax credits, tax deductions, tax exemptions, and rebates. *Ownership incentives* are recurring annual or periodic financial benefits that accrue to PEV owners regardless of use and include exemptions from (or reductions in) registration taxes or fees, weight surcharges, environmental taxes, or vehicle inspections. *Use incentives* are ongoing financial benefits realized by driving a PEV and include exemptions from motor fuel taxes, reduced roadway taxes or tolls, and discounted or free PEV charging or parking. *PEV infrastructure incentives* are one-time financial benefits for deploying PEV charging stations and include tax credits, rebates, or other subsidies. A variety of these incentives have been used throughout the United States and in other countries. Educating consumers on all the incentives is challenging, and some confusion results because incentives vary by location and often come and go without much warning.

The primary consumer incentive offered by the U.S. federal government is a purchase incentive in the form of a tax credit. The tax credit amount varies depending on the capacity of the battery in the vehicles and will be phased out at the beginning of the second calendar quarter after the manufacturer produces 200,000 eligible PEVs as counted from January 1, 2010.[12] To claim the credit, consumers who purchase a PEV must have sufficient tax liability and will not see the benefit until they file an annual tax return. For consumers who lease, the leasing company typically claims the credit and reflects the credit in the monthly lease rate, so leasers essentially see the benefit of the tax credit at the point of sale. Although the PEV tax credits are analogous to the HEV and diesel-vehicle tax credits in the 1990s and 2000s, which have since expired, the notable differences are that the tax credit for most PEVs is much higher than the HEV and diesel credits. Because more people lease PEVs than purchase them, a higher fraction of PEV drivers see the benefits of the credit sooner; therefore, the effect of the PEV tax credits could be greater than the effect of the now expired HEV and diesel credits. However, a recent study found that 94.5 percent of survey respondents (adult drivers from the general public in 21 major U.S. cities) were not aware of PEV incentives and suggests that the effectiveness of the PEV credits could be enhanced through greater consumer awareness and education (Krause et al. 2013).

The U.S. state governments have offered a variety of financial incentives (see Table 7-2). The DOE Alternative Fuels Data Center maintains a database that provides a comprehensive listing of state incentives.[13] Several states have offered purchase incentives in the form of tax credits in addition to the one offered by the federal government. The monetary amount varies from state to state; for example, as of August 2014, available tax credits ranged from $605 in Utah to up to $6,000 in Colorado. The tax credits have also varied over time; many have been reduced or recently expired. The method for calculating the credit varies from state to state; some states simply calculate it on the basis of purchase price, and others use battery capacity and purchase price to determine the amount. Several states have also used sales-tax exemptions or rebates to make the effect of the purchase incentive more immediate for those who choose to buy rather than lease. Some of these purchase incentives are restricted to certain types of PEVs. For example, the sales-tax exemptions in Washington and New Jersey are restricted to BEVs, and the rebate in Illinois is restricted to BEVs and range-extended PHEVs. California, however, provides rebates to BEVs and PHEVs, although the amount differs ($2,500 for BEVs and $1,500 for PHEVs in 2014). Using rebates and sales-tax exemptions is consistent with recent research that compares the effectiveness of HEV tax credits, sales-tax exemptions, and rebates and finds that the sales-tax exemptions and rebates appear to be more effective than tax credits possibly because of their immediacy, transparency, and simplicity (Chandra et al. 2010; Gallagher and Muehlegger 2011).

State governments have also used ownership and use incentives to promote PEV deployment. The most common have been exemptions from registration fees or vehicle inspections and reduced roadway taxes or tolls. Local governments have also offered discounted or free PEV charging or parking. States have also provided financial incentives for installing PEV charging stations so that consumers will be

[12] The federal tax credit is $2,500 for PEVs that have battery capacities below 5 kWh. For PEVs that have larger battery capacities, the credit is set at $2,500 plus $417 times the amount that the battery capacity exceeds 5 kWh, up to a maximum of $7,500.

[13] See U.S. Department of Energy, "State Laws and Incentives," http://www.afdc.energy.gov/laws/state.

BOX 7-2 Financial Incentives

Tax credits and *tax deductions* are taken at the end of the tax reporting period. They act to lower the final year-end taxes owed to federal or state governments. *Tax credits* are considered more desirable because they directly offset a taxpayer's liability in the exact amount of the credit. For example, if the end-of-year tax liability for a person was $18,000, a tax credit of $5,000 would directly lower the total taxes owed by that same amount, to $13,000.

In contrast, *tax deductions* reduce the amount of reported income that is subject to taxation, rather than directly offsetting taxes owed. If persons had taxable income of $60,000 (taxed at 25 percent), they would owe $15,000. If they took a $5,000 tax deduction, their taxable income would be reduced by that amount, to $55,000, which in turn would lower their tax liability by $1,250 to $13,750. Tax deductions are often subject to rules that limit the amounts that can be deducted or that restrict higher-income taxpayers from taking the full deduction.

As financial incentives, many *tax credits* are available to all persons who file a tax return whereas *tax deductions* are available only to those persons who file a tax return that itemizes deductions. Studies show that in the United States fewer than 50 percent of all federal tax returns claim itemized deductions (Prante 2007).

Tax exemptions are recognized at the time of a transaction (for example, at the point of sale) or during a regular tax reporting period (for example, vehicle registration renewal process). By exempting an entire asset or activity from taxation, the financial benefits are often realized immediately, such as with a sales tax exemption on a vehicle purchase. Tax exemptions are not usually subject to income-based qualifications or limitations, as is the case with many tax deductions.

Rebates provided by the government can take several forms, depending on their structuring. The key distinguishing feature of a rebate is that it is earned (and often processed) at the time of a qualifying purchase. Some rebate programs require an individual to submit proof of a qualifying purchase directly to the government to receive a rebate check; other rebates are provided to the seller of qualifying goods or services so that the total purchase price to the consumer can be reduced in an equal amount. However structured, both consumers and sellers tend to prefer rebates over tax credits, deductions, or exemptions because the financial benefits are immediately realized at the time of the purchase transaction, regardless of tax rates and method of tax filing.

A *fee-bate* is a method of taxing or applying a surcharge or fee on certain activities or classes of assets that are deemed to have undesirable social attributes to generate sufficient revenue to provide direct rebates for other activities or assets that are deemed to be more desirable. Because this section is more narrowly focused on the types of financial incentives that can be provided rather than the method of funding those incentives, a fee-bate system and rebates are treated in the same way because they both result in a rebate.

A *subsidy* is a more general term used to describe methods for government-provided financial assistance. A subsidy can take the common form of tax credits, deductions, exemptions, or rebates; or, a subsidy can include direct government grants, lower than market rate loans, loan guarantees, or myriad other ways for government to provide financial support.

sure that they will be able to charge their vehicle away from home. The most common and popular nonfinancial incentive offered by the states has been access to restricted lanes, such as bus-only, high-occupancy-vehicle, and high-occupancy-toll lanes. That incentive has been used by several states over the years to promote adoption of PEVs (and HEVs).

Other countries have used incentives similar to those used by the United States, as summarized in Table 7-2. The most popular have been purchase incentives in the form of tax exemptions or rebates, ownership incentives in the form of exemptions or reductions in registration or ownership taxes or fees, and use incentives in the form of reduced roadway taxes or tolls. Some of the financial incentives have been substantial. For example, Norway offers substantial tax breaks (no purchase tax, no annual registration tax, and no value-added tax) that amount to about $11,000 over the vehicle lifetime, or about $1,400 per year (Doyle and Adomaitis 2013). Commuters also do not pay road tolls, which are worth $1,400 annually, and receive free parking, which is worth $5,000. They are also permitted to drive in bus lanes and have access to free public charging at over 450 locations in Oslo (Doyle and Adomaitis 2013). Another example is the Netherlands, which had financial incentives that equaled as much as 85 percent of the vehicle price, although these have been reduced. It is important to note that the financial incentives in the Netherlands are particularly important because electricity prices are so high that the consumer's incentive to use electricity as a fuel is small.

One interesting purchase rebate program is the one offered by the Clean Energy Vehicle Promotion Program in Japan. It is notable because it has a clear sunset, the rebate level declines every year on the basis of a preset formula, and the rebate amount financed by the government depends on whether vehicle manufacturers meet a preset annual price target (see Figure 7-1). The administering agency, the Ministry of Economy, Trade and Industry (METI 2013) calculates

TABLE 7-2 Incentives for Plug-in Electric Vehicles (PEVs) by Country and State

Type of Incentive	Location
Financial Incentives	
Purchase Incentives—one-time financial benefit earned by purchase of PEV	
Tax credits or deductions (realized only on filing tax return)	U.S. federal government United States: Colorado, Georgia, Louisiana, South Carolina, Utah Other countries: Australia, Austria, Belgium, Israel
Tax exemptions or rebates (realized at the point of sale)	United States: California, District of Columbia, Illinois, Massachusetts, New Jersey, Pennsylvania, Texas, Washington Other countries: Canada (Ontario [for leased vehicles], British Columbia [purchased or leased], and Quebec [leased]), China, Estonia, France, Iceland, Ireland, Israel, Japan, Luxembourg, the Netherlands, Norway, Portugal, Spain, Sweden, U.K.
Ownership Incentives—recurring annual or periodic financial benefit that accrues to PEV owners, regardless of use	
Exemption from or reduction in registration or ownership taxes or fees	United States: Arizona, Connecticut, District of Columbia, Illinois, Maryland Other countries: Australia (Victoria), Austria, Belgium (Flanders), Denmark, Finland, Germany, Greece, Ireland, Italy, Japan Latvia, the Netherlands, Norway, Romania, Sweden, Switzerland (varies by region), U.K.
Exemption from or reduction in weight surcharges (collected annually at time of registration or renewal)	United States: Colorado Other countries: Japan
Exemption from environmental taxes	Other countries: Denmark
Exemption from vehicle inspection	United States: Arizona, California, Colorado, Connecticut, Delaware, District of Columbia, Georgia, Idaho, Illinois, Maryland, Massachusetts, Missouri, Nevada, New Jersey, New Mexico, New York, Oregon, Pennsylvania, Tennessee, Utah, Virginia, Washington, Wisconsin
Use Incentives—on-going financial benefits realized by driving a PEV	
Exemption from motor fuel taxes	United States: North Carolina, Utah, Wisconsin Other countries: European Union, Japan, Norway
Reduced roadway taxes or tolls	United States: California, New Jersey, New York Other countries: Austria, Czech Republic, Japan, the Netherlands, Norway, Switzerland (varies by region), U.K.
Discounted or free PEV charging	United States: Arizona, California, Delaware, Georgia, Illinois, Indiana, Kentucky, Maryland, Michigan, Minnesota, Nevada, Virginia Other countries: Japan, Norway
Discounted or free PEV parking	United States: Free parking available at some airports, such as Long Beach Airport; at parking garages in some states and localities, such as Nevada, Sacramento, and Santa Monica; and other locations, often with free charging Other countries: Denmark, Iceland, Norway
PEV Infrastructure Incentives—one-time financial benefit for deploying PEV charging stations	
Tax credit or rebate for installing PEV charging station	United States (individual and business): Arizona, California, Florida, Georgia, Illinois, Indiana, Louisiana, Maryland, Michigan, New York, Oklahoma, Oregon
PEV charging infrastructure deployment subsidies	United States (individual and business): California, Colorado, Connecticut, District of Columbia, Florida, Illinois, Maryland, Massachusetts, Nebraska, New Mexico, Ohio, Texas, Utah, Washington Other countries: Canada, European Union, Israel, Japan, Korea, Norway
Nonfinancial Incentives	
Use Incentives—on-going special privileges granted to PEV drivers	
Access to restricted lanes, such as bus-only, high-occupancy-vehicle, and high-occupancy-toll lanes	United States: Arizona, California, District of Columbia, Florida, Georgia, Hawaii, Maryland, New York, North Carolina, Tennessee, Utah, Virginia Other countries: the Netherlands, Norway
Reserved parking for PEVs	United States: Arizona, California, Hawaii, Washington

SOURCES: Based on data from Gallagher and Steenblick (2013); Brand et al. (2013); Beltramello (2012); Morrow et al. (2010); Tesla (2013); DOE (2014b); Doyle and Adomaitis (2013); EV Norway (2014); Mock and Yang (2014); IEA (2013); Jin et al. (2014).

an annual price target by assuming a linear decline between a base price in 2012 and a long-term target price in 2016. To encourage vehicle manufacturers to reduce their sales prices every year, the government provides 100 percent of the rebate if the manufacturer meets the annual target price but subsidizes only about 67 percent of the rebate if the manufacturer exceeds the annual price target.

It is difficult to draw firm conclusions about the experiences with incentive programs in other countries given the cultural, political, and geographical differences. However, countries with substantial financial incentives for PEVs, such as Norway and the Netherlands, have seen a high rate of PEV adoption. Those with little or no financial incentives for PEVs—most notably Germany, which has not offered consumer incentives and has relied on demonstration programs in four major regions—have experienced minimal sales. Financial incentives, however, are not working everywhere, most notably in China, where there has been tepid consumer uptake despite the substantial financial incentives offered. One early analysis of that puzzling situation concludes that Chinese consumers are more concerned about vehicle performance than cost at this stage (Zhang et al. 2013). Further information on the international experience is provided in Appendix C.

There has been little academic research about the effectiveness of fiscal incentives in stimulating the adoption of PEVs. However, a greater body of evidence now exists regarding fiscal incentives and HEVs. Overall, that literature suggests that financial incentives do motivate consumers to purchase more fuel-efficient vehicles (see, for example, Huang 2010; Sallee 2011; Ozaki and Sevastyanova 2011). In general, it also seems that the more immediate the incentive, the more effective it is at persuading consumers to purchase the more fuel-efficient vehicle. Sales-tax exemptions or reductions and rebates at the state level have been associated much more strongly with consumer adoption, presumably due to their immediacy and ease of transaction. The federal cash-for-clunkers program, for example, offered a purchase rebate and resulted in strong consumer response to the immediate subsidy (Huang 2010).

Finding: Given the research on fiscal incentives and HEVs, the effectiveness of the federal income tax credit to motivate consumers to purchase PEVs would be enhanced by converting it into a rebate at the point of sale.

Finding: The U.S. state and local governments offer a variety of financial and nonfinancial incentives; there appears, however, to be a lack of research to indicate which incentives might be the most effective at encouraging PEV deployment.

Finding: The many state incentives that differ in monetary value, restrictions, and calculation methods make it challenging to educate consumers on the incentives that are available to them and emphasize the need for a clear, up-to-date source of information for consumers.

Finding: Overall, the experience worldwide demonstrates that substantial financial incentives are effective at motivating consumers to adopt PEVs.

PRICE OF CONVENTIONAL TRANSPORTATION FUELS AS AN INCENTIVE OR A DISINCENTIVE FOR THE ADOPTION OF PLUG-IN ELECTRIC VEHICLES

High gasoline prices motivate consumers to drive less and to purchase a more fuel-efficient vehicle, at least for some time after prices rise noticeably. As noted by Diamond (2009, p. 982),

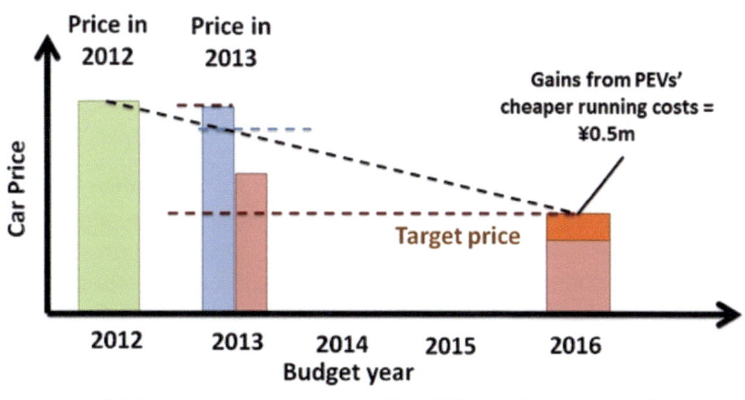

FIGURE 7-1 Japan's clean energy vehicles promotion program. If a PEV's price exceeds the dashed black line, the government subsidizes two-thirds of the difference. If a PEV's price is below the dashed black line, the government subsidizes 100 percent of the difference. SOURCE: Based on data from METI (2013).

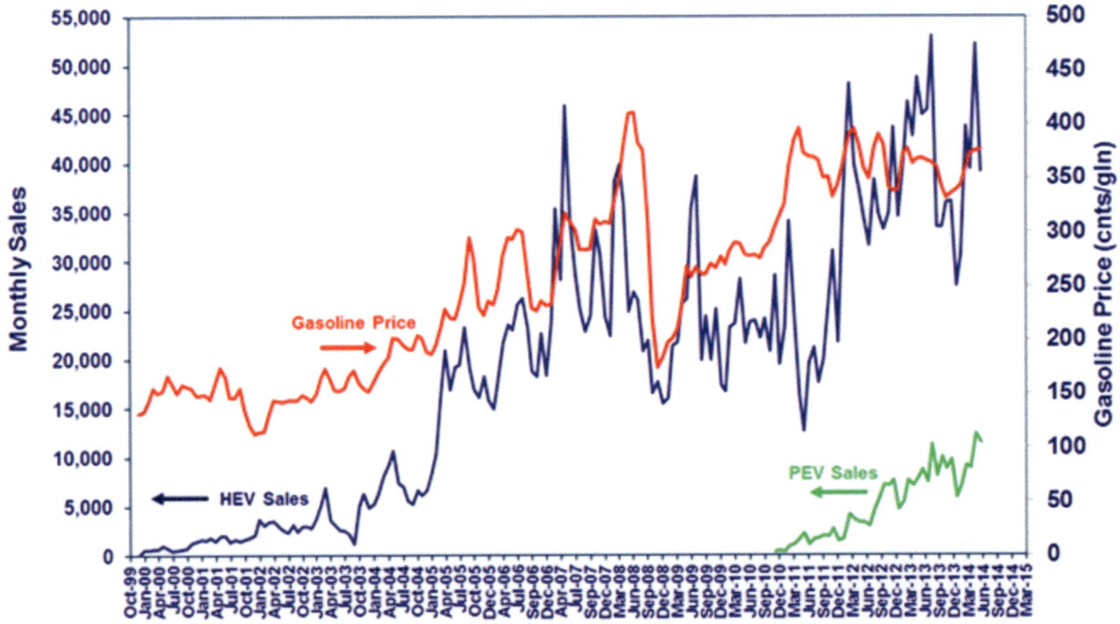

FIGURE 7-2 U.S. HEV and PEV sales overlaid with U.S. gasoline prices. NOTE: HEV, hybrid electric vehicle; PEV, plug-in electric vehicle. SOURCE: ANL (2014).

> Gasoline prices serve as the most visible signal for consumers to think about fuel savings and fuel economy, so it is reasonable that relatively minor variations in gasoline prices could lead to significant changes in adoption patterns, particularly for people in the market for a new car as gas prices rise or fall.

Rapid increases in gasoline prices that increased adoption rates for HEVs provide some support for that assertion (see Figure 7-2). By the same token, low gasoline prices create a disincentive for PEV adoption. They reduce the savings in fuel costs that a consumer would realize by owning a PEV and could make the cumulative cost of PEV ownership appear less attractive than the same cost of a conventional ICE vehicle.

As of January 2015, U.S. gasoline prices were less than half of those in most European countries, including Belgium, France, Germany, Italy, the Netherlands, and the U.K. (EIA 2015). The higher gasoline prices in Europe and Asia are mostly due to considerably higher gasoline taxes, which more than double the price of gasoline per gallon. Accordingly, numerous studies in the United States and elsewhere have concluded that taxes on conventional transportation fuels that substantially raise the gasoline price create an incentive for consumers to purchase more fuel-efficient vehicles and to drive fewer conventional-vehicle miles (Diamond 2009; Morrow et al. 2010; Small 2012; Burke and Nishitateno 2013).

Broader market-based policies like carbon taxes and cap-and-trade regimes theoretically could create a disincentive for the use of conventional vehicles and an incentive for the use of PEVs. It is important to note, however, that the carbon taxes applied by a few countries (Denmark, Norway, Sweden, and Switzerland) and by the Canadian province of British Columbia on transportation fuels do not strongly affect the prices of petroleum fuels because the carbon content of gasoline and diesel fuels is much less than that of coal. California's low-carbon fuel standard, which imposed a compliance cost of \$13 per ton CO_2 emissions, was assessed by Yeh and Witcover (2012) and was found to add one-tenth of a penny per gallon to the cost of gasoline in 2012. Thus, although carbon taxes and cap-and-trade regimes might be the most effective methods for reducing GHG emissions, they might not provide a meaningful incentive to purchase a PEV.

PAST INCENTIVES ON OTHER ALTERNATIVE VEHICLES AND FUELS

Over the past few decades, a number of federal and state policy initiatives have been implemented to stimulate the deployment of alternative vehicles and fuels. Air quality, climate change, and energy security concerns have motivated the initiatives. The primary alternative vehicles and fuels that have been considered in the light-duty fleet include HEVs, PEVs, and hydrogen FCVs, and methanol, ethanol, natural gas, propane, and biodiesel for use as fuels in conventional ICE vehicles.[14] Key laws and regulations that are aimed di-

[14] For the purposes of this report, the focus will be primarily on the lessons learned from alternative vehicles and fuels in light-duty vehicle applications.

rectly at alternative fuels or that provide incentives for alternative-fuel vehicles include the Alternative Motor Fuels Act of 1988, the Clean Air Act of 1990, the Energy Policy Act of 1992, the California ZEV program (originally adopted in 1990), the Renewable Fuel Standard (part of the Energy Policy Act of 2005 and the Energy Independence and Security Act of 2007),[15] and the GHG and CAFE standards.

The public sector approach to developing, promoting, or deploying alternative-fuel vehicles includes (1) research and development, (2) demonstration projects, (3) fleet deployment, (4) niche market development, (5) public-private partnerships, and (6) various policy and financial incentives. The general approach has been based on the supposition that alternative-fuel vehicles would need to be subsidized until the point where the life-cycle costs of the vehicles and fuel would become competitive with those of gasoline-fueled vehicles; market forces would thereafter operate without subsidies, leading to broad deployment (NRC 2008, 2010a, 2013b).

In some cases in which advancements in technology were needed, government and private-sector funding of research and development led to a technology push. DOE partnered with the private sector on vehicle technologies and fuels through such activities as the Partnership for a New Generation of Vehicles (PNGV), the FreedomCAR and Fuel Partnership, and, currently, the U.S. Driving Research and Innovation for Vehicle Efficiency and Energy Sustainability (U.S. DRIVE) Partnership (NAS/NAE/NRC 2009; NRC 2008, 2010b, 2013a). Extensive work on bioenergy crops and technologies for their conversion into ethanol and, now, "drop-in" biofuels[16] has also been funded (NRC 2011b).

To create a market pull for some of the technologies under development, various tax or policy mandates aimed at stimulating market demand were initiated. Financial incentives were especially needed for new technologies that were projected to be more expensive than the incumbent conventional technologies, at least in the initial and transitional phases. For example, tax credits for certain types of vehicles and tax breaks for certain fuels were put into the tax code to stimulate the adoption of the new, more costly technologies, at least for a period of time until increased production and economies of scale drove costs down to the point where the new technologies would be competitive in the marketplace. It was also thought that adoption by fleets and promotion in niche markets would, in many cases, help with this transition period by increasing sales and production and thus driving down costs, but studies evaluating the programs do not provide a clear picture of whether that strategy is useful (Leiby and Rubin 2004; McNutt and Rodgers 2004; Robertson and Beard 2004; Hwang 2009; NRC 2010b, 2013a; Greene 2012).

[15] Pub. L.110-140, 121 Stat. 1758. USC § 17001.

[16] The term *drop-in biofuel* refers to the conversion of biomass into fuels, such as gasoline and diesel, that are compatible with, and can be "dropped into," the current fueling infrastructure. This approach would avoid overcoming the barriers to developing and investing in the infrastructure necessary for a separate fueling system, such as would be required for fuels, such as ethanol, methanol, natural gas, and hydrogen.

Methanol

There was great interest in methanol (MeOH) as a potential motor vehicle fuel in the late 1980s and early 1990s for a number of reasons (API/WRI 1990). Its high octane content could be used in some ICEs with a higher compression ratio to improve efficiency; it could also result in lower emissions than conventional gasoline-fueled vehicles and thus lead to improved air quality and better public health. From a national security point of view, it could displace imported petroleum if it was produced from biomass or from natural gas. The conversion of global sources of remote natural gas that were of low economic value was envisioned as an approach to diversify the U.S. global supply chain for light-duty vehicle fuels and replace petroleum.

Given extensive experience with the use of MeOH and ICEs in the world of competitive racing, the development of MeOH-powered vehicles did not necessitate fundamental technology breakthroughs and such vehicles were developed for the market that could operate on either high levels of methanol, such as 85 percent MeOH and 15 percent gasoline (referred to as M85), or on any combination between M0 and M85. These flexible-fuel vehicles (FFVs) were sold in the marketplace and represented a strategy for overcoming the chicken-and-egg problem of attracting investment in a methanol-fueling infrastructure before there are enough MeOH-fueled vehicles on the road and the lack of demand for such vehicles until an extensive fueling infrastructure is in place. It was anticipated that with MeOH FFVs, the vehicle owner could use the existing gasoline infrastructure while a MeOH-fueling infrastructure was built in response to vehicles deployed for fleets, incentives were implemented, and a business case for fuel investors became viable.

Despite subsidies, a broad M85 infrastructure never materialized. Furthermore, the continued improvement of gasoline-powered vehicles along with the development of reformulated gasoline resulted in gasoline-powered vehicles that could achieve the same or better emission performance as promised by MeOH and essentially eliminated the need for MeOH-fueled vehicles.

Ethanol

Similar to MeOH, ethanol (EtOH) is a fuel with a high octane content and one that could be produced from a variety of domestic resources, although it has an energy density per unit volume only two-thirds that of gasoline.[17] It continues to be of interest with a focus on, as with methanol, the development of FFVs that can operate on mixtures from 0 percent EtOH and

[17] The lower volumetric energy density results in a lower miles-per-gallon fuel efficiency compared with gasoline and, all else being equal, will require more frequent refueling.

100 percent gasoline to 85 percent EtOH and 15 percent gasoline (E85) to address the chicken-and-egg problem. EtOH was originally used as a gasoline additive during summer months to reduce air pollutant emissions. Later, the EtOH program was viewed as a means of displacing petroleum and using domestic resources for EtOH production. Vehicle manufacturers were given credits in CAFE regulations, and this led to a significant production of ethanol-capable FFVs. Although the program might have been a successful transition strategy for ultimately replacing gasoline with ethanol, the reality was that most of the ethanol-capable FFVs used little, if any, ethanol (NRC 2002).

In addition, the federal government invested a great deal of research and development in the development of nonfood crops (such as species of trees and grasses) that could be grown on energy plantations and whose cellulose could serve as a feedstock for conversion technologies to produce EtOH (NAS/NAE/NRC 2009; NRC 2011b). It was envisioned that, if successful, renewable fuel production system would have low net GHG emissions and enhance energy security and would not compete with land for the production of food. The development of cost-effective cellulosic-based EtOH technologies that can compete with gasoline has proven more difficult to achieve than anticipated. But there is ongoing demonstration and development of such biomass conversion technologies, and it remains to be seen how much this alternative fuel will contribute to the U.S. transportation fuel supply.

As with MeOH, an extensive system of fueling stations supplying E85 has yet to emerge, even though there are millions of EtOH FFVs on the road (most of which use gasoline) and the U.S. Congress mandated the use of a certain amount of EtOH through the Renewable Fuel Standard (NRC 2011b). To date, most EtOH is produced from corn or sugar cane and is used as a renewable-fuel replacement for petroleum with up to 10 percent EtOH blended into gasoline. In some places, mostly the Midwest, E15 can be sold and used in light-duty vehicles with a model year 2001 or later. Increasing the percentage for conventional vehicles has received opposition from some quarters because of the potentially deleterious effects of ethanol on engine components, particularly marine engines, although this is not an issue for FFVs. However, the aggressive policies and subsidies have led to about 7 percent replacement of gasoline-energy use in light-duty vehicles from less than 1 percent in 2000, and this demonstrates that sustained efforts by the federal government can have demonstrable effects in the market (Gruenspecht 2013).

Compressed Natural Gas

Another alternative-vehicle system that garnered interest in the 1990s and one that is also used worldwide are vehicles using compressed natural gas (CNG). They offer air quality advantages in urban areas and can be fueled from domestic sources of natural gas, which seemed plentiful and cheap in the 1990s as it has again in the past few years. Most of the vehicles developed were dedicated CNG vehicles that avoided the extra vehicle cost and complexity that would be needed for a dual-fuel vehicle, although some dual-fuel natural gas vehicles were offered in the market in 2012-2013, stimulated by low natural gas prices and projections of future plentiful reserves and associated low prices.

CNG is typically stored onboard the vehicle at 3,600 psi and requires high-pressure fueling stations. Incentives and mandates were provided in the 1990s to bring the vehicles to market, but because of the need for high-pressure fueling stations and bulky storage tanks on the vehicle and the shorter driving range compared with comparable gasoline-powered vehicles, they tended to be used in fleets where the vehicles returned to a central station at the end of each day and could be refueled. They were somewhat more expensive than comparable gasoline-powered vehicles, trunk space was somewhat compromised, driving range was shorter, and an extensive refueling infrastructure was not, and still is not, available. Consumers did not embrace CNG vehicles, and these vehicles have not moved beyond the niche fleet markets.

Fuel-Cell Vehicles

Another major alternative vehicle and fuel technology that has been promoted and developed to varying degrees by the public and private sector is the hydrogen-fueled FCV. It uses on-board hydrogen in a fuel cell to produce electric power to drive the vehicle. Because its only emission is water vapor, it is classified by California as a ZEV. The federal government and the private sector have provided substantial funding for research and development, for vehicle demonstrations, and for parts of the needed hydrogen infrastructure (NRC 2010b, 2013a). There has been significant technical progress and promise of driving ranges and fueling times comparable with those of conventional vehicles, but they are still a work in progress. Cost-effective production of hydrogen, deploying the necessary hydrogen infrastructure, and overcoming the chicken-and-egg barriers remain formidable challenges for these vehicles. Some vehicle manufacturers have indicated that such vehicles will be available for the market in the 2015-2016 time frame, and Hyundai began leasing a fuel-cell vehicle in California in 2014. However, the higher costs of these vehicles compared with conventional vehicles will be substantial, and thus their deployment will require subsidies and other new technologies to overcome the initial cost barrier (NRC 2008, 2013b).

Hybrid Electric Vehicles

In 1999, HEVs were introduced into the U.S. automotive market (ANL 2014). A federal income tax incentive for HEVs existed between 2000 and 2010. The original tax incentives provided a tax deduction of up to $2,000, but the Energy Policy Act of 2005 increased it to a maximum of $3,400 and converted it into a tax credit. The deductions and credits were the maximums granted for the most fuel-efficient vehicles. HEVs with lesser fuel-economy received more modest tax credits. The tax credits were available for the first 60,000 vehicles sold by a manufacturer, after which time the tax credits would expire. In addition to the federal income tax credits, states offered a wide array of other consumer incentives, including income tax credits, sales-tax reductions or exemptions, access to high-occupancy-vehicle (HOV) lanes, reduced registration fees, and exemptions from emissions testing, similar to the incentives now offered for PEVs.

The effectiveness of the purchase incentives for HEVs has been extensively studied in the United States and elsewhere. As noted above, the most important finding from the literature is that, immediate purchase incentives (such as a sales-tax exemption or instant rebate) are more effective than tax credits or deductions because consumers appear to focus on up-front price and highly discount long-term cost savings (Diamond 2009; Chandra et al. 2010; Gallagher and Muehlegger 2011). With immediate incentives, buyers do not have to wonder whether they will qualify for the credit when they file taxes in the next year or estimate its value given their income bracket. With immediate incentives, the purchase price can be adjusted at the time of sale. A study of Canadian experience with tax rebates for HEVs, which were established at the point of sale, found that they were highly effective (Chandra et al. 2010).

Lessons Learned from Past Incentive Programs

The past incentive programs for alternative-fuel technologies indicate that the market for advanced technology adoption needs to be cultivated to progress beyond early adopters. Sustained efforts and economic incentives that create a profitable business case, however, can have demonstrable effects. The ethanol example is one where the regulatory mandate was successful at advancing an alternative-fuel technology; the percentage of ethanol in the domestic gasoline supply by volume increased from less than 1 percent in 2000 to 10 percent in 2011. Using fleets to encourage mainstream adoption does not appear to be particularly effective (for example, in the case of CNG vehicles). Cost reduction and technology advances will continue to evolve as product volumes increase, but in the meantime, financial incentives are needed to make a technology more cost-competitive. The hybrid example, with a U.S. adoption rate still below 4 percent, shows that even with financial incentives and substantial technology advances, moving the deployment from successful regional and niche markets to mainstream adoption remains a challenge.

RECOMMENDATIONS

Because there has not been any extensive research on regulatory or incentive programs that promote PEV adoption, it was difficult for the committee to determine which incentives would be the most effective and should be pursued. However, on the basis of its review of the barriers to PEV adoption and current and past federal and state incentive programs, the committee offers the following recommendations:

Recommendation: Federal financial incentives to purchase PEVs should continue to be provided beyond the current production volume limit as manufacturers and consumers experiment with and learn about the new technology. The federal government should re-evaluate the case for incentives after a suitable period, such as 5 years. Its re-evaluation should consider advancements in vehicle technology and progress in reducing production costs, total costs of ownership, and emissions of PEVs, HEVs, and ICE vehicles.

Recommendation: Given the research on effectiveness of purchase incentives, the federal government should consider converting the tax credit to a point-of-sale rebate.

Recommendation: Given the sparse research on incentives other than financial purchase incentives, research should be conducted on the variety of consumer incentives that are (or have been) offered by states and local governments to determine which, if any, have proven effective in promoting PEV deployment.

REFERENCES

Al-Alawi, B.M., and T.H. Bradley. 2014. Analysis of corporate average fuel economy regulation compliance scenarios inclusive of plug in hybrid vehicles. Applied Energy 113:1323-1337.

ANL (Argonne National Laboratory). 2014. "Light Duty Electric Drive Vehicles Monthly Sales Updates." http://www.transportation.anl.gov/technology_analysis/edrive_vehicle_monthly_sales.html. Accessed August 20, 2014.

API/WRI (American Petroleum Institute and World Resources Institute). 1990. Methanol as an Alternative Fuel Choice: An Assessment. Washington, DC: International Energy Program, Foreign Policy Institute, the Paul H. Nitze School of Advanced International Studies, Johns Hopkins University, Washington, DC.

Beltramello, A. 2012. "Market Development for Green Cars." OECD Green Growth Papers. No. 2012-03. Paris: OECD Publishing.

Brand, C., J. Anable, and M. Tran. 2013. Accelerating the transformation to a low carbon passenger transport system: The role of car purchase taxes, feebates, road taxes, and scrappage incentives in the UK. Transportation Research Part A: Policy and Practice 49: 132-148.

Burke, P., and S. Nishitateno. 2013. Gasoline prices, gasoline consumption, and new vehicle fuel economy: Evidence from a large sample of countries. Energy Economics 36: 363-370.

CARB. 2013. Final Regulation Order §1962.2 Zero Emission Vehicle Standards for 2018 and Subsequent Model Year Passenger Cars, Light-Duty Trucks, and Medium-Duty Vehicles. http://www.arb.ca.gov/msprog/zevprog/2011zevreg/attachb3.pdf.

Chandra, A., S. Gulati, and M. Kandlikar. 2010. Green drivers or free riders? An analysis of tax rebates for hybrid vehicles. Journal of Environmental Economics and Management 60(2): 78-93.

Diamond, D. 2009. The impact of government incentives for hybrid-electric vehicles: Evidence from US states. Energy Policy 37(3): 972-983.

DOE (U.S. Department of Energy). 2011. One Million Electric Vehicles by 2015: February 2011 Status Report. http://www1.eere.energy.gov/vehiclesandfuels/pdfs/1_million_electric_vehicles_rpt.pdf.

DOE. 2014a. "Vehicle Cost Calculator." Alternative Fuels Data Center. http://www.afdc.energy.gov/calc/.

DOE. 2014b. "State Laws and Incentives." Alternative Fuels Data Center. http://www.afdc.energy.gov/laws/state.

Doyle, A., and N. Adomaitis. 2013. "Norway Shows the Way with Electric Cars, but at What Cost? Reuters, March 13. http://www.reuters.com/article/2013/03/13/us-cars-norway-idUSBRE92C0K020130313.

Edwards, A. 2013. "The PEV Customer: How to Overcome Potential Sales Barriers." Presentation to the Committee on Overcoming Barriers to Electric-Vehicle Deployment, Washington, DC, December 3.

EIA (U.S. Energy Information Administration). 2015. "Retail Premium Gasoline Prices, Selected Countries." http://www.eia.gov/countries/prices/gasolinewithtax.cfm.

EPA/NHTSA (U.S. Environmental Protection Agency and National Highway Traffic Safety Administration). 2012a. 2017 and Later Model Year Light-Duty Vehicle Greenhouse Gas Emissions and Corporate Average Fuel Economy Standards. Federal Register 77(199): October. http://www.gpo.gov/fdsys/pkg/FR-2012-10-15/pdf/2012-21972.pdf.

EPA/NHTSA. 2012b. Joint Technical Support Document: Final Rulemaking for 2017-2025 Light-Duty Vehicle Greenhouse Gas Emission Standards and Corporate Average Fuel Economy Standards. Report No. EPA-420-R-12-901. Washington, DC.

EV Norway. 2014. "Powered by Nature: A Guide to Norway, the World's Leading EV Market." http://www.evnorway.no/. Accessed January 31, 2015.

Gallagher, K.S., and M. Muehlegger. 2011. Giving green to get green? Incentives and consumer adoption of hybrid vehicle technology. Journal of Environmental Economics and Management 61(1): 1-15.

Gallagher, K.S., and R. Steenblick. 2013. Technology-Push and Demand-Pull Policies to Support Electric Propulsion. OECD Report COM/TAD/ENV/JWPTE, Paris, France.

Greene, D.L. 2012. "Alternative Fuels: What Have We Learned?" Presentation at ARPA-e Transportation Behavior and New Technology Workshop, Berkeley, CA, March 16. http://arpa-e.energy.gov/sites/default/files/documents/files/Alternative%20Fuels%20-%20What%20Have%20We%20Learned.pdf.

Gruenspecht, H. 2013. "Biofuels in the United States: Context and Outlook." Presented at the Biofuels Workshop, Institute of Medicine, National Academy of Sciences, Washington, DC, January 24. http://www.eia.gov/pressroom/presentations/howard_01242013.pdf.

Huang, E. 2010. "Do Public Subsidies Sell Green Cars? Evidence from the U.S. 'Cash for Clunkers' Program." Discussion Paper #2010-17. Harvard Kennedy School, Belfer Center, Cambridge, MA. December.

Hwang, R. 2009. "U.S. Alternative Fuels Policies: Lessons Learned and Future Directions." Presentation at Senate Briefing on Alternative Fuels. Washington, DC, February 6.

IEA (International Energy Agency). 2013. Global EV Outlook: Understanding the Electric Vehicle Landscape to 2020. http://www.iea.org/publications/globalevoutlook_2013.pdf.

IRS (Internal Revenue Service). 2009. "Qualified Vehicles Acquired after 12-31-2009." Last reviewed or updated November 17, 2014. http://www.irs.gov/Businesses/Qualified-Vehicles-Acquired-after-12-31-2009.

Jin, L., S. Searle, and N. Lutsey. 2014. "Evaluation of State-Level U.S. Electric Vehicle Incentives." White Paper presented to the International Council on Clean Air Transportation. http://www.theicct.org/sites/default/files/publications/ICCT_state-EV-incentives_20141030.pdf.

Keddie, E. 2013. "California's Zero Emission Vehicle (ZEV) Regulation." Presentation to the Committee on Overcoming Barriers to Electric-Vehicle Deployment, Irvine, CA, May 8.

Krause, R.M., S.R. Carley, B.W. Lanea, and J.D. Graham. 2013. Perception and reality: Public knowledge of plug-in electric vehicles in 21 U.S. cities. Energy Policy 63: 433-440.

Leiby, P., and J. Rubin. 2004. "Understanding the Transition to New Fuels and Vehicles: Lessons Learned from Analysis and Experience of Alternative Fuel and Hybrid Vehicles." In The Hydrogen Energy Transition: Moving Towards the Post-Petroleum Age in Transportation, edited by D. Sperling and J. Cannon, 191-212. Waltham, MA: Elsevier Academic Press.

McNutt, B., and D. Rodgers. 2004. "Lessons Learned from 15 Years of Alternative Fuels Experience—1988 to 2003." In The Hydrogen Energy Transition: Moving Towards the Post-Petroleum Age in Transportation, edited by D. Sperling and J. Cannon, 165-179. Waltham, MA: Elsevier Academic Press.

METI (Ministry of Economy, Trade, and Industry). 2013. "Our Policy about Promoting EVs." Presentation to the

Committee on Overcoming Barriers to Electric-Vehicle Deployment, Tokyo, Japan, December 9.

Mock, P., and Z. Yang. 2014. "Driving Electrification: A Global Comparison of Fiscal Incentive Policy for Electric Vehicles." White paper submitted to the International Council of Clean Transportation, Washington, DC, May. http://www.theicct.org/sites/default/files/publications/ICCT_EV-fiscal-incentives_20140506.pdf.

Morrow, W.R., K.S. Gallagher, G. Collantes, and H. Lee. 2010. Analysis of policies to reduce oil consumption and greenhouse-gas emissions from the U.S. transportation sector. Energy Policy 38(3): 1305-1320.

NAS/NAE/NRC (National Academy of Sciences, National Academy of Engineering, and National Research Council). 2009. America's Energy Future: Technology and Transformation. Washington, DC: The National Academies Press.

NRC (National Research Council). 2002. Effectiveness and Impact of Corporate Average Fuel Economy (CAFÉ) Standards. Washington, DC: The National Academies Press.

NRC. 2008. Transitions to Alternative Transportation Technologies—A Focus on Hydrogen. Washington, DC: The National Academies Press.

NRC. 2010a. Transitions to Alternative Transportation Technologies: Plug-in Hybrid Electric Vehicles. Washington, DC: The National Academies Press.

NRC. 2010b. Review of the Research Program of the FreedomCAR and Fuel Partnership, Third Report. Washington, DC: The National Academies Press.

NRC. 2011a. Assessment of Fuel Economy Technologies for Light-Duty Vehicles. Washington, DC: The National Academies Press.

NRC. 2011b. Renewable Fuel Standard: Potential Economic and Environmental Effects of U.S. Biofuel Policy. Washington, DC: The National Academies Press.

NRC. 2013a. Review of the Research Program of the U.S. DRIVE Partnership, Fourth Report. Washington, DC: The National Academies Press.

NRC. 2013b. Transitions to Alternative Vehicles and Fuels. Washington, DC: The National Academies Press.

Ohnsman, A. 2013. "Tesla Tops California Green-Car Credit Sales in Past Year." Bloomberg Businessweek, October 16. http://www.businessweek.com/news/2013-10-16/musk-s-tesla-tops-california-green-car-credit-sales-in-past-year.

Ozaki, R., and K. Sevastyanova. 2011. Going hybrid: An analysis of consumer purchase motivations. Energy Policy 39(5): 2217-2227.

Prante, G. 2007. "Most Americans Don't Itemize on Their Tax Returns." Tax Foundation, July. http://taxfoundation.org/article/most-americans-dont-itemize-their-tax-returns.

Robertson, B., and L.K. Beard. 2004. "Lessons Learned in the Deployment of Alternative Fueled Vehicles." In The Hydrogen Energy Transition: Moving Towards the Post-Petroleum Age in Transportation, edited by D. Sperling and J. Cannon, 181-190. Waltham, MA: Elsevier Academic Press.

RTI/UMTRI (Research Triangle Institute and University of Michigan Transportation Research Institute). 2009. Automobile Industry Retail Price Equivalent and Indirect Cost Multipliers: Report Prepared for the U.S. Environmental Protection Agency. RTI Project Number 0211577.002.004. Washington, DC.

Sallee, J.M. 2011. The surprising incidence of tax credits for consumers. American Economic Journal: Economic Policy 3(2): 189-219.

Small, K. 2012. Energy policies for passenger motor vehicles. Transportation Research Part A: Policy and Practice 46(6): 874-889.

State ZEV Programs. 2013. "State Zero-Emission Vehicle Programs—Memorandum of Understanding." Signed October 24, 2013. http://arb.ca.gov/newsrel/2013/8s_zev_mou.pdf.

Tellis, G. 2013. "Toyota's Gamble on the Prius." Financial Times Management. Financial Times, March 4. http://www.ft.com/intl/cms/s/0/146ad23c-7230-11e2-89fb-00144feab49a.html#axzz3IJ1hL64u.

Tesla. 2013. "Electric Vehicle Incentives Around the World." http://www.teslamotors.com/incentives/japan. Accessed February 14, 2013.

Yeh, S., and J. Witcover. 2012. Status Review of California's Low Carbon Fuel Standard. Series UCD-ITS-RP-12-33. University of California, Davis. August.

ZEV Program Implementation Task Force. 2014. Multi-State ZEV Action Plan. March. http://www.nescaum.org/documents/multi-state-zev-action-plan.pdf/.

Zhang, X., K. Wang, Y. Hao, J. Fan, and Y. Wei, 2013. The impact of government policy on preference for NEVs: The evidence from China. Energy Policy 61: 382-393.

Appendixes

A

Biographical Information on the Committee on Overcoming Barriers to Electric-Vehicle Deployment

JOHN G. KASSAKIAN, *Chair*, is a professor of electrical engineering and former director of the Massachusetts Institute of Technology (MIT) Laboratory for Electromagnetic and Electronic Systems. His expertise is in the use of electronics for the control and conversion of electric energy, industrial and utility applications of power electronics, electronic manufacturing technologies, and automotive electric and electronic systems. Before joining the MIT faculty, he served in the U.S. Navy. Dr. Kassakian is on the boards of directors of a number of companies and has held numerous positions with the Institute of Electrical and Electronic Engineers (IEEE), including as founding president of the IEEE Power Electronics Society. He is a member of the National Academy of Engineering, a fellow of IEEE, and a recipient of the IEEE William E. Newell Award for Outstanding Achievements in Power Electronics (1987), the IEEE Centennial Medal (1984), and the IEEE Power Electronics Society's Distinguished Service Award (1998). He is a coauthor of the textbook *Principles of Power Electronics* and has served on a number of National Research Council committees, including the Electric Power/Energy Systems Engineering Peer Committee, the Committee on Assessment of Solid State Lighting, the Committee on Review of the 21st Century Truck Partnership Phase 2, the Committee on Review of the Research Program of the Partnership for a New Generation of Vehicles, and the Committee on Review of the FreedomCAR and Fuel Research Program. He has an ScD in electrical engineering from MIT.

DAVID L. BODDE is an engineering professor and senior fellow in Clemson University's International Center for Automotive Research. Before joining Clemson University, Dr. Bodde held the Charles N. Kimball Chair in Technology and Innovation at the University of Missouri–Kansas City. He serves on the boards of directors of several energy and technology companies, including Great Plains Energy and the Commerce Funds. His executive experience includes vice president of Midwest Research Institute and president of MRI Ventures, assistant director of the Congressional Budget Office, and deputy assistant secretary in the U.S. Department of Energy. He holds a doctorate in business administration from Harvard University, MS degrees in nuclear engineering and management from the Massachusetts Institute of Technology, and a BS from the U.S. Military Academy. He served in the Army in Vietnam.

JEFF DOYLE is a principal with D'Artagnan Consulting, LLP, a firm that specializes in the use of advanced technologies to achieve important transportation policy and finance objectives. Prior to that, Mr. Doyle served as director of public-private partnerships for the Washington State Department of Transportation (WSDOT), where he oversaw a program that develops partnerships with the private sector to advance transportation projects, programs, and policies. His office implemented the West Coast Electric Highway, a partnership project that provided the nation with its first border-to-border network of fast-charging stations for electric vehicles. Mr. Doyle also served as co-chair of Washington's Plug-in Vehicle Task Force, as a member of the Puget Sound Regional Council Electric Vehicle Infrastructure Advisory Committee, and as a member of the National Cooperative Highway Research Program Project Panel 20-83 (04), "Effects of Changing Transportation Energy Supplies and Alternative Fuel Sources on Transportation." Other public-private partnership projects include redeveloping public ferry terminals, providing transit-oriented development with advanced traveler-information systems at state-owned park-and-ride facilities, and implementing alternative finance and funding mechanisms for transportation infrastructure development and maintenance. Prior to joining WSDOT, Mr. Doyle served as staff director and senior legal counsel to the Transportation Committee in the Washington legislature, where his work focused on transportation policy, finance, and freight mobility issues. He is a member of the Washington State Bar Association and serves on the Supervisory Committee of a state-chartered credit union in Washington. Mr. Doyle earned a BA in political science from Western Washington University and a JD from Seattle University.

GERALD GABRIELSE is Leverett Professor of Physics at Harvard University. His previous positions include assistant and associate professor, University of Washington, and chair of the Harvard Department of Physics. His research focuses on making accurate measurements of the electron magnetic moment and the fine structure constant and on the precise laser spectroscopy of helium. Dr. Gabrielse also leads the International Antihydrogen TRAP (ATRAP) Collaboration, whose goal is accurate laser spectroscopy with trapped antihydrogen atoms. His many awards and prizes include fellowship of the American Physical Society, the Davisson-Germer prize of the American Physical Society, the Humboldt Research Award (Germany, 2005), and the Tomassoni Award (Italy, 2008). Harvard University awarded him its George Ledlie Research Prize and its Levenson Teaching Prize. His hundreds of outside lectures include a Källén Lecture (Sweden), a Poincaré Lecture (France), a Faraday Lecture (Cambridge, U.K.), a Schrödinger Lecture (Austria), a Zachariasen Lecture (University of Chicago), and a Rosenthal Lecture (Yale). He is a member of the National Academy of Sciences and has participated on many National Research Council committees, including the Committee on Review of the U.S. DRIVE Research Program, Phase 4, and the Committee on Review of the FreedomCAR and Fuel Research Program, Phase 3. He has a BS from Calvin College and an MS and a PhD in physics from the University of Chicago.

KELLY SIMS GALLAGHER (*committee member until June 2014*) is an associate professor of energy and environmental policy of the Fletcher School, Tufts University. She directs the Energy, Climate, and Innovation research program of the Center for International Environment and Resource Policy. She is also a senior associate and a member of the Board of Directors of the Belfer Center for Science and International Affairs of Harvard University, where she previously directed the Energy Technology Innovation Policy research group. Broadly, she focuses on energy and climate policy in the United States and China. She is particularly interested in the role of policy in spurring the development and deployment of cleaner and more efficient energy technologies domestically and internationally. She speaks Spanish and basic Mandarin Chinese and is a member of the Council on Foreign Relations. She is the author of *China Shifts Gears: Automakers, Oil, Pollution, and Development* (MIT Press, 2006), editor of *Acting in Time on Energy Policy* (Brookings Institution Press, 2009), and author of numerous academic articles and policy reports. A Truman Scholar, she has an MA in Law and Diplomacy and a PhD in international affairs from the Fletcher School and an AB from Occidental College.

ROLAND HWANG is the transportation program director for the Natural Resources Defense Council (NRDC) and works on sustainable transportation policies. He is an expert on clean vehicle and fuels technologies and was a member of the Technology Assessment and Economics Panel of the Intergovernmental Panel on Climate Change, which won the 2007 Nobel Peace Prize. Mr. Hwang serves or has served on numerous advisory panels, including the California Plug-in Electric Vehicle Collaborative, the National Academy of Sciences Committee on Fuel Economy, the U.S. EPA Mobile Source Technical Review Subcommittee, the California Air Resources Board's Alternative and Renewable Fuels and Vehicles program, the California Hydrogen Highway Network Advisory Panel, the Automotive X Prize, and the Western Governors' Association Transportation Fuels for the Future Initiative. Before joining NRDC, he was the director of the Union of Concerned Scientists transportation program. He has also worked for the U.S. Department of Energy at Lawrence Berkeley National Laboratory and the California Air Resources Board as an air pollution engineer and was involved in forecasting residential and industrial energy demand, permitting of hazardous waste incinerators, and evaluating toxic air emissions from landfills. He is currently on the National Research Council Committee on Fuel Economy of Light-Duty Vehicles, Phase 2. Mr. Hwang has an MS in mechanical engineering from the University of California, Davis, and a master's degree in public policy from the University of California, Berkeley.

PETER ISARD is a consultant on economic policy issues and held various managerial positions with the International Monetary Fund (IMF) from 1985 to 2008, primarily in the Research Department. Dr. Isard played a lead role in helping Lithuania design an economic transformation program in 1991 and 1992 and spent the 2002-2003 academic year at the University of Maryland. He retired in June 2008 as deputy director of the IMF Institute, the department that provides training on economic policy making for member-country officials. Before joining the IMF in 1985, he spent 1970 in the Research Department of the IMF, taught at Washington University in St. Louis in 1971-1972, held research and management positions at the Federal Reserve Board from 1972 to 1985, and spent a year during 1979 and 1980 at the Bank for International Settlements. Dr. Isard is the author of numerous articles in academic journals, primarily on exchange rates and monetary policy strategies. He is also the author of two books—*Exchange Rate Economics* (1995) and *Globalization and the International Financial System* (2005)—and editor of several others. He has an undergraduate degree in mathematics from the Massachusetts Institute of Technology and a PhD in economics from Stanford University.

LINOS JACOVIDES is professor of electrical engineering at Michigan State University. From 1998 to 2007, he served as director of Delphi Research Laboratories. Dr. Jacovides joined General Motors Research and Development in 1967 and became department head of electrical engineering in 1985. His research was in the interactions between power electronics and electric machines in electric vehicles and locomotives. He later transitioned to Delphi with a group of researchers from General Motors to set up the Delphi Research Laboratories. He is a member of the National Academy of Engineering and has served on numerous National Research Council committees, including the National Cooperative Highway Research Program's Panel on Effects of Changing Transportation Energy Supplies and Alternative Fuel Sources on State Departments of Transportation, the Committees on Assessment of Technologies for Improving Light-Duty Vehicle Fuel Economy (Phases 1 and 2), the Committee on Review of the U.S. Drive Research Program, Phase 4, the Committee on Electric Vehicle Controls and Unintended Acceleration, and the Committee on Review of the Freedom-CAR and Fuel Research Program, Phase 3. Dr. Jacovides is a fellow of IEEE and the Society of Automotive Engineers (SAE) International and served as president of the Industry Applications Society of IEEE in 1990. He received a BS in electrical engineering and an MS in machine theory from the University of Glasgow and a PhD in generator control systems from the Imperial College, University of London.

ULRIC KWAN is a senior managing consultant with IBM, concentrating on helping utilities around the globe enable customer-oriented systems of engagement. Previously, he worked for Pacific Gas and Electric, where he was a leader in the electric vehicle and demand response fields. In this role, he was responsible for the development of business and regulatory justification for utility involvement in electric-vehicle deployments, integration of electric vehicles and customers into the electricity market, creation of a long-term contract between an automaker and utility for demand response, and development of the first integrated point-of-sale rate calculator for electric-vehicle customers. Earlier, he worked at Siemens as an energy engineer and LCG Consulting as an electricity wholesale market consultant and forecaster. Mr. Kwan has participated in numerous regulatory proceedings at the California and federal electricity regulatory bodies on how to integrate customers into the wholesale market and serves or has served on numerous advisory panels, including the North American Electric Reliability Corporation (NERC) advisory team for demand response, the California Plug-in Electric Vehicle Collaborative, and the California Electric Transportation Coalition. Mr. Kwan has a BS and MS in mechanical engineering from the University of Calgary and Stanford University, respectively.

REBECCA LINDLAND is a senior fellow with the King Abdullah Petroleum Studies and Research Center and leads its work on transportation policy, technology, and consumer demand. She was formerly the director of research for IHS Automotive where she was responsible for evaluating and assessing vehicle manufacturers that participate in the U.S. and Canada marketplaces. She has a particular interest in how manufacturers' decisions reflect consumer values. As a member of IHS Automotive, Ms Lindland was frequently quoted in the media, including the *New York Times*, *Business Week*, *Reuters*, *Bloomberg News*, the *Los Angeles Times*, the *Detroit News*, the *Detroit Free Press*, the *Wall Street Journal*, and National Public Radio, for her coverage of new product launches and the balance sheet conditions of manufacturers and brands. Prior to her work at IHS, Ms Lindland worked at AlliedSignal in Rumford, Rhode Island, where she forecasted products, such as Bendix brakes. A life-long automotive enthusiast, she began her career as a staff accountant with Mercedes-Benz Credit Corporation in Norwalk, Connecticut. Ms Lindland holds a double major in accounting and business administration from Gordon College, Wenham, Massachusetts. She is a former board member of the Society of Automotive Analysts, the International Motor Press Association, and Motor Press Guild and was accepted into Strathmore's 2001 *Who's Who in American Business*.

RALPH D. MASIELLO is the senior vice president and innovation director of DNVGL, Inc. In recent years, his focus has been on electric market and transmission operator business models and systems, including cost-benefit analyses of paradigms for models, systems, and operations. He has also developed technology and strategic plans for market operators and automation and smart grid roadmaps for several independent system operators. His current interests include the market and utility applications of advanced storage devices for ancillary markets, reliability, and energy economics; the grid integration of electric vehicles; and the development of advanced building-to-grid concepts. He has provided expert testimony before Congress on metering systems and market operations and cosigned a Supreme Court amicus curiae brief on transmission access and native load service. He was recently appointed to the Department of Energy Electricity Advisory Council. Dr. Masiello is a fellow of the IEEE and has served as chairman of Power System Engineering, as chairman of Power Industry Computing Applications, on the editorial board of *IEEE Proceedings* and on the advisory board of *IEEE Spectrum* magazine. He is the winner of the 2009 IEEE PES Concordia award for power system analysis and is a member of the National Academy of Engineers. He received his PhD from the Massachusetts Institute of Technology in electrical engineering.

JAKKI MOHR is the Regents Professor of Marketing at the University of Montana–Missoula (UM). An international expert and innovator in marketing of high-technology products and services, she has achieved international acclaim for *Marketing of High-Technology Products and Innovations* (coauthor with S. Sengupta and S. Slater, with European and India/Southeast Asia editions and translations into Chinese, Portuguese, and Korean). Motivated by the desire to apply the promise of new technologies to solve social and global problems, Dr. Mohr has provided training to companies and universities worldwide in strategic market planning to commercialize innovation. She has received numerous teaching awards—including the Outstanding Marketing Teacher

Award (presented by the Academy of Marketing Science), the Carnegie Foundation CASE Professor of the Year, and the Most Inspirational Teacher of the Year Award at the University of Montana—and the Distinguished Scholar Award, the John Ruffatto Memorial Award, and the Dennison Presidential Faculty Award for Distinguished Accomplishment. Dr. Mohr served as a Fulbright senior specialist in Montevideo, Uruguay. Her research has received national awards and has been published in the *Journal of Marketing*, the *Strategic Management Journal*, the *Journal of the Academy of Marketing Science*, the *Journal of Public Policy and Marketing*, and others. In research sponsored by the Marketing Science Institute, she studies how companies use biomimicry (innovations inspired by nature that are based on underlying biologic mechanisms) to solve technical and engineering challenges, the basis of her TEDxSanDiego talk in 2011. Before joining UM in fall 1997, Dr. Mohr was an assistant professor at the University of Colorado Boulder (1989-1997). Before beginning her academic career, she worked in Silicon Valley in advertising for Hewlett-Packard's Personal Computer Group and TeleVideo Systems. Dr. Mohr received her PhD from University of Wisconsin–Madison.

MELISSA SCHILLING is a professor of management and organizations at New York University Stern School of Business. Dr. Schilling teaches strategic management, corporate strategy, and technology and innovation management. She is widely recognized as an expert in innovation and strategy in high-technology industries. Her textbook, *Strategic Management of Technological Innovation* (now in its fourth edition), is the top innovation-strategy text in the world and is available in seven languages. Her research in innovation and strategy has earned her such awards as the National Science Foundation's CAREER Award and the Best Paper in Management Science and Organization Science for 2007 Award. Her research has appeared in leading academic journals, such as the *Academy of Management Journal*, the *Academy of Management Review*, *Management Science, Organization Science, the Strategic Management Journal*, and the *Journal of Economics and Management Strategy and Research Policy*. She sits on the editorial review boards of *Organization Science* and *Strategic Organization*. Dr. Schilling received her BS in business administration from the University of Colorado, Boulder, and her PhD in strategic management from the University of Washington.

RICHARD TABORS is president of Across the Charles and is director of the Utility of the Future Project at the MIT Energy Initiative. Until July 2012, he was vice president of Charles River Associates (CRA) in the Energy & Environment Practice. He founded the engineering-economics consulting firm of Tabors Caramanis & Associates (TCA) in 1989 to provide economic, regulatory, and financial analytic support to the restructuring of the U.S. and international electric power industry. TCA was sold to CRA in 2004. He was a researcher and member of the faculty at Harvard University from 1970 to 1976 and was at Massachusetts Institute of Technology as a senior lecturer in technology management and policy and a research director in power systems from 1976 to 2004. He is a visiting professor of electrical engineering at the University of Strathclyde in Glasgow. His research and development activities at MIT led to his authorship or coauthorship of over 80 articles and books, including *Spot Pricing of Electricity*, on which the economic restructuring of the electric utility wholesale and retail markets is based. Dr. Tabors continues his directing and consulting activities in regulation, litigation, and asset evaluation in the power industry with a focus on development of future platforms and pricing structure of the smart grid. He received a BA in biology from Dartmouth and an MS and a PhD in geography and economics from the Maxwell School of Syracuse University.

TOM TURRENTINE is director of the California Energy Commission's Plug-in Hybrid and Electric Vehicle Research Center at the Institute of Transportation Studies, University of California, Davis. For the last 20 years, Dr. Turrentine has been researching consumer response to alternative fuels, vehicle technologies, road systems, and policies that have environmental benefits. His most recent work includes multiyear projects to study consumer use of plug-in electric vehicles, including the BMW Mini E, Prius PHEV conversions, the Nissan Leaf, GM Volt, PHEV pickups, and specially designed energy-feedback displays in vehicles. He and his researchers are studying BEV and PHEV driver travel patterns and use of infrastructure and are developing planning tools to advise on deployment of infrastructure and optimal ways to integrate plug-in electric vehicles into California's grid. He and his team wrote "Taking Charge," a plan for California to develop a PEV market, which is the blueprint for the California PEV Collaborative. He holds a PhD in anthropology.

B

Meetings and Presentations

FIRST COMMITTEE MEETING
October 28-29, 2012

EV Everywhere: Overview and Status
Patrick B. Davis, Program Manager, Vehicle Technologies Program, U.S. Department of Energy (DOE)

EV Everywhere Grand Challenge: Charging Infrastructure Enabling Flexible EV Design
Lee Slezak, Technology Manager, Vehicle Systems, Vehicle Technologies Program, DOE

DOE AVTA: The EV Project and Other Light-Duty Electric Drive Vehicle Activities
James Francfort, Principal Investigator, Advanced Vehicle Testing Activity, Idaho National Laboratory

SECOND COMMITTEE MEETING
December 17-19, 2012

Charging Infrastructure Needs
Marcus Alexander, Manager, Vehicle Systems Analysis, Electric Power Research Institute (EPRI)

General Motors: National Research Council
Britta K. Gross, Director, Advanced Vehicle Commercialization Policy, General Motors

Overcoming Barriers to Deployment of Electric Vehicles
Mike Tamor, Executive Technical Leader, Energy Systems and Sustainability, Ford Motor Company

Overcoming Barriers to Electric-Vehicle Deployment: Barriers to Deployment, an OEM Perspective
Joseph Thompson, Project Manager-Technology Planning, Nissan

The Electrification Coalition: Revolutionizing Transportation and Achieving Energy Security
Jonna Hamilton, Vice President for Policy, Electrification Coalition

Electric Vehicle Charging Services
Richard Lowenthal, Founder and CTO, ChargePoint

The Complete Electric Vehicle Charging Solution
Michael Krauthamer, Director, Mid-Atlantic Region, eVgo

Better Place Update
Jason Wolf, Vice President for North America, Better Place

The DOE Vehicle Technologies Analysis Toolbox and EV Everywhere Target-Setting
Jacob Ward, Vehicle Technologies Analysis Manager, DOE

The Need for Public Investments to Support the Plug-in Electric Vehicle Market
Nick Nigro, Manager, Transportation Initiatives, Center for Climate and Energy Solutions

Research Insights from the Nation's Highest Residential Concentration of Electric Vehicles
Brewster McCracken, President and CEO, Pecan Street Inc.

Electric Vehicle Initiatives in the Houston-Galveston Region
Allison Carr, Air Quality Planner, Houston-Galveston Area Clean Cities Coalition

The EV Project Deployment Barriers
Donald Karner, ECOtality North America

New Models of Mobility and EV Deployment
Jack Hidary, Global EV Leader, Hertz

Electric Vehicle Infrastructure Demonstration Projects: Lessons Learned
Rick Durst, Transportation Electrification Project Manager, Portland General Electric

THIRD COMMITTEE MEETING
January 25-26, 2013

No open session presentations were held during this meeting.

FOURTH COMMITTEE MEETING
May 8-9, 2013

California's Zero-Emission-Vehicle (ZEV) Regulation
Elise Keddie, Manager, Zero Emission Vehicle Implementation Section, California Air Resources Board

Electric-Vehicle Deployment: A Long-Term Perspective
Chuck Shulock, President, Shulock Consulting

Perspective on the Electrification of the Automotive Fleet: The Prius and Beyond
Toyota Motor Corporation

Consumer Behavior and Attitudes Concerning PEV Adoption
Ed Kim, Vice President, Industry Analysis, AutoPacific

Selling Plug-in Electric Vehicles
Paul Scott, EV Specialist, Downtown LA Nissan

San Diego Gas & Electric Plug-in Electric Vehicle Landscape
John H. Holmes, Research & Development, Asset Management & Smart Grid Projects, San Diego Gas & Electric

FIFTH COMMITTEE MEETING
August 13-14, 2013

DOE Electric Vehicle Activities Update
Jake Ward, Program Analyst, Vehicles Technology Program, DOE

Vehicle Choice Modeling for Advanced Technology and Electric Vehicles
David Greene, Corporate Fellow, National Transportation Research Center, Oak Ridge National Laboratory

Local Barriers to Plug-in Electric Vehicle (PEV) Deployment
Katie Drye, Transportation Project Manager, Advanced Energy

Workplace Charging Challenge: Part of the EV Everywhere Grand Challenge
Sarah Olexsak, Energy Project Specialist, DOE

Workplace Electric Vehicle Charging
Ali Ahmed, Senior Manager, Workplace Resources, Global Energy Management and Sustainability, Cisco Systems, Inc.

Workplace Charging Programs – Nissan
Tracy Woodard, Senior Director for Government Affairs, Nissan

EV Charging at Lynda.com
Dana Jennings, Facilities Supervisor, Lynda.com, Inc.

Panel Discussion on Workplace Charging
Katie Drye, Sarah Olexsak, Ali Ahmed, Tracy Woodard, Dana Jennings

Technical, Manufacturing, and Market Issues Associated with xEV Batteries
Suresh Sriramulu, Vice President, Battery Technology, TIAX LLC

Consumers' Thoughts, Attitudes, and Potential Acceptance of Electric Vehicles
Chris Travell, Vice President of Automotive Research, Maritz Research

SIXTH COMMITTEE MEETING
December 3-4, 2013

The PEV Customer: How to Overcome Potential Sales Barriers
Alexander Edwards, President, Strategic Vision

Panel Discussion: Dealer Perspective on Plug-in Electric Vehicles
Tammy Darvish, Vice President, DARCARS Automotive Group

Neil Kopit, Director of Marketing, Criswell Automotive, Gaithersburg, MD

Greg Brown (via teleconference), General Manager, Serra Chevrolet, Southfield, MI

Doug Greenhaus, Chief Regulatory Counsel, Environment, Health and Safety, National Automobile Dealers Association

PEV Deployment in the Defense Department: Barriers and Strategies
Camron Gorguinpour, Executive Director, Plug-in Electric-Vehicle Program, Department of Defense

Massachusetts Electric Vehicle Roadmap
Mark Sylvia, Commissioner, Massachusetts Department of Energy Resources

FEDEX Experience
Russ Musgrove, Managing Director, FedEx Express

Frito-Lay Experience
Steve Hanson, Fleet Sustainability Manager, Frito-Lay

Panel Discussion on Fleet Deployment
Camron Gorguinpour, Mark Sylvia, Russ Musgrove, and Steve Hanson

SEVENTH COMMITTEE MEETING
February 25-26, 2014

The Future of Automobile Battery Recycling
Linda Gaines, Transportation Systems Analyst, Center for Transportation Research, Argonne National Laboratory

Li-Ion Technology Evolution for xEVs: How Far and How Fast?
Menahem Anderman (via WebEx), President, Advanced Automotive Batteries

Electric Vehicle Charging Infrastructure Usage Observed in Large-Scale Charging Infrastructure Demonstrations
John Smart, Electric Vehicle Test Engineer, Energy Storage & Transportation Systems, Idaho National Laboratory

What Electric-Vehicle Drivers Want in a Charging Network (and What They Actually Need)
Michael Nicholas, Institute of Transportation Studies, University of California, Davis

Understanding Electric Vehicle Market Barriers: An Automotive Manufacturer's Perspective
William P. Chernicoff, Manager, Energy and Environmental Research, Toyota Motors North America, Inc.

EV Infrastructure Financing Solutions
John Rhow, Kleiner Perkins

Reporting on Site Visits to Japan
Roland Hwang, Member, Committee on Overcoming Barriers to Electric-Vehicle Deployment, Transportation Program Director, Natural Resources Defense Council

EIGHTH COMMITTEE MEETING
May 6-7, 2014

Stationary Wireless Charging of PEVs: Near-Term Barriers
John Miller, JNJ Miller plc

Car2Go: Electric Vehicles and Car Sharing
Mike Cully, U.S. Regional Manager, Car2Go

DOE Vehicle Electrification Activities
Patrick Davis, Program Manager, Vehicle Technologies, DOE

Reporting on Site Visits to Europe
Jeff Doyle, Member, Committee on Overcoming Barriers to Electric-Vehicle Deployment, Director Public/Private Partnerships, Washington State Department of Transportation

NINTH COMMITTEE MEETING
July 16-17, 2014

No open session presentations were held during this meeting.

TENTH COMMITTEE MEETING
October 23-24, 2014

No open session presentations were held during this meeting.

C

International Incentives

This appendix provides some information about the incentive programs in Japan, France, Norway, Germany, the Netherlands, and China.

JAPAN

In fiscal year (FY) 2013, the Japanese government offered rebates for 18 different makes and models of plug-in electric vehicles (PEVs) available in Japan. For the Nissan Leaf, the FY 2013 rebate was ¥780,000 (about $7,800) based on a 2013 target price of ¥2,850,000 (about $28,500) (METI 2013).[1] The committee predicts that the rebates will decline to ¥520,000 (about $5,200) in FY 2014, ¥260,000 (about $2,600) in FY 2015, and zero in FY 2016. In addition to the rebates, PEV purchasers are also exempt from the vehicle acquisition tax (about 5 percent of the purchase price) and from the vehicle weight or tonnage tax (Nelson and Tanabe 2013). The acquisition tax is waived through March 2015, and the weight tax is waived through April 2015 (Tesla 2013). The vehicle weight or tonnage tax exemption is applicable once, at the time of the first mandatory inspection, which occurs 3 years after the vehicle purchase. PEV owners also enjoy a substantial discount on the annual automobile tax, which can otherwise range from ¥29,500 to ¥111,000, depending on the vehicle's engine displacement. Finally, some prefectures and cities offer additional incentives at time of purchase.

FRANCE

In 2007, France introduced a fee-bate (bonus-malus) system for vehicle purchases based on the carbon dioxide (CO_2) emissions of the vehicle. The policy levies a fee depending on the CO_2 emission performance of the vehicle ranging from €150 to €8,000 and provides a rebate ranging from €150 to €6,300.[2] The dealer can advance the bonus at the point of sale to reduce the purchase price directly. PEVs qualify for the highest bonus of €6,300. The bonus-malus system generated deficits in its first few years (2008-2010) owing to unexpectedly high demand for the lower-CO_2 emitting vehicles but led to substantial reductions in the CO_2 emissions of new vehicles sold in France (Beltramello 2012). Average new light-duty vehicle CO_2 per kilometer moved from being the fourth lowest to the lowest in the European Union since the program started in 2007 (Brand et al. 2013). The bonus-malus system is periodically updated, with the most recent revision having become effective in January 2014.

The bonus-malus system appears to be an effective consumer incentive. According to the French government, the French market for PEVs and hybrid electric vehicles (HEVs) represented 3.1 percent of the global passenger vehicle market in France. Compared with 2012, sales of PEVs increased by 50 percent and sales of HEVs increased by 60 percent. In total, 8,779 PEVs were registered in France in 2013. Sales increased by more than 50 percent compared with the 5,663 vehicles registered in 2012.

NORWAY

The government of Norway has made a firm commitment to battery electric vehicles (BEVs), motivated in part by the desire to reduce the greenhouse gas (GHG) emissions of its transportation fleet. Because almost 100 percent of Norway's electricity is generated from hydroelectric power, a transition to BEVs would decarbonize the passenger vehicle fleet almost entirely. Forty percent of Norway's GHG emissions currently come from the transportation sector, and 60 percent of those come from road transport (Deshayes 2011).

According to a recent study of an incentive scheme scheduled to last through 2017 (Doyle and Adomaitis 2013), the Norwegian government provides tax breaks of up to $11,000 over the lifetime of a PEV, or about $1,400 per year. The tax breaks include no purchase tax, no annual registration tax, and no value-added tax (VAT) (Doyle and Adomaitis 2013). As part of the scheme, commuters do not pay road

[1] The fiscal year for the national Japanese budget cycle runs from June to May.

[2] For more specific break-downs on the bonus-malus system, see: http://www.developpement-durable.gouv.fr/Bonus-Malus-2014.html.

tolls, worth $1,400 annually, and they receive free parking worth $5,000. PEVs are permitted in bus lanes and have access to free public charging at 466 parking spots in Oslo (Doyle and Adomaitis 2013).

As of the beginning of 2013, PEV sales accounted for 3 percent of total passenger car sales, a much higher fraction than in most countries. A total of 12,000 PEVs had been sold in Norway as of 2013, with about half in the Oslo region (Ingram 2013a). Nonetheless, some (for example, Doyle and Adomaitis 2013) have criticized the incentive program because it encourages families to purchase a PEV as a second car and rely on their gasoline- or diesel-powered vehicle for longer-range trips. However, even if that is the practice, families might be driving more electric miles during the course of everyday life. Although the programs could prove to be an environmental benefit, Norway might not be able to sustain such a financial commitment. It spends about $13,600 in tax incentives to reduce CO_2 emissions by just one tonne. This cost is much higher than the prevailing price of CO_2 on the European Union emissions trading market (Ingram 2013b).

GERMANY

Germany does not currently offer consumer incentives and is instead relying on a demonstration program in four major regions. German vehicle manufacturers are investing heavily in hydrogen fuel-cell vehicles (FCVs), and, other than BMW, they have been slow to embrace PEVs.

THE NETHERLANDS

The Netherlands has extensive consumer incentives for PEVs and at one time these incentives equaled as much as 85 percent of the price of a new plug-in hybrid electric vehicle (PHEV), although they have since been scaled back. Unsurprisingly, the Netherlands has become a hot market for PEV manufacturers and is Tesla's second biggest market besides the United States after Norway. The Dutch government is especially motivated to support BEVs because most of the larger cities in the Netherlands experience severe urban air pollution. Municipal governments are also keen to reduce urban noise, especially in the evenings, and find that noise reduction from BEV taxis and delivery vans greatly improves the quality of city life (Nissan 2012). The Dutch government also views BEV deployment as consistent with its climate change goals and strategy. Not having significant domestic vehicle production, there is little resistance to importing BEVs from abroad.

The tax incentive structure is unique among all the countries examined because corporate buyers overwhelmingly dominate the Dutch new-vehicle market, and most new vehicles are bought by firms for their employees. Employees must pay income tax (bijtelling) for vehicles received from their employers. For example, 25 percent of the value of a new vehicle is added to an employee's personal income, and then he or she must pay income tax on the total. The bijtelling tax is assessed on the basis of grams of CO_2 per kilometer, and for high-emitting vehicles, the tax rate is 25 percent. For BEVs, the bijtelling tax rate is 4 percent. BEV buyers also enjoy a purchase tax incentive, whereby through 2017 they pay no tax for vehicles with low CO_2 emissions and are exempt from a vehicle-use tax, which is normally based on weight and kilometers driven. Employees are therefore motivated to encourage their employers to buy them BEVs. The federal government is also providing a purchase subsidy for BEV taxis and delivery vans used in urban areas to help cope with urban air pollution and noise. Amsterdam, Arnhem, The Hague, Rotterdam, and Utrecht add an additional purchase subsidy for taxis and delivery vans (€5,000) and trucks (€40,000) and are particularly motivated as no new construction may occur in the city until air pollution has been reduced (Dutch Ministry of Economic Affairs 2014). The incentives are especially important to consumers because Dutch electricity prices are high (€0.28/kWh), so the consumer incentive to use electricity as a fuel is minimal.

CHINA

Beginning in 2006, China made a major push toward PEVs. Given China's heavy reliance on coal to generate electricity, the main environmental benefits for the country could be cleaner air in some cities and a reduction in noise pollution. However, according to a recent analysis by Ji et al. (2012), replacing gasoline vehicles with PEVs in China with its current electricity supply mix will result in higher CO_2 emissions and increased mortality risk from $PM_{2.5}$ (particulate matter less than 2.5 micrometers in diameter) in most Chinese cities. In any case, the Chinese government views a shift to PEVs to be beneficial to China's energy security. China is a net importer of coal and its current reserve-to-production ratio of coal is only 31 years (BP 2013). The energy security benefits are therefore not apparent. As of March 2013, there were about 28,000 PEVs registered in China, of which about 80 percent were public buses.

As of 2010, there were 135 million electric bicycles in China (Jie and Hagiwara 2013).[3] China is already the largest electric bicycle producer and consumer, accounting for about 90 percent of the global market. The Chinese government research and development program for clean, light-duty vehicles initially focused almost equally on FCVs, BEVs, and PHEVs. In China's Five-Year Plan (2006-2010), however, the government's emphasis shifted strongly to BEVs.

The Chinese central government has subsidized the deployment of PEVs since 2009. Some local governments in 25 pilot cities also provided subsidies on top of the central government subsidies discussed below, mostly to support the

[3] As of 2008, 970 invention patents had been applied for through the State Intellectual Property Organization (SIPO) based on the research of the Chinese government's Energy-Saving and New Energy Vehicle Programme (Ouyang 2009).

purchase of public transportation vehicles, such as buses and locally stated-owned taxis. It has been alleged that some local governments have imposed "buy local" provisions so that the local PEV firms benefit at the expense of PEV companies elsewhere in China and around the world (Zeng 2013).

The Chinese government allowed six cities to experiment with subsidies to individual consumers who purchased PEVs starting in 2013.[4] In those cities, the local government is allowed to provide purchase incentives, and the central government will also provide up to RMB 50,000 (about $8,000) for the purchase of a PHEV and RMB 60,000 (about $9,600) for the purchase of a BEV. Beijing has announced that it will also subsidize BEVs at a rate of RMB 60,000 (about $9,600), while Shanghai will provide a subsidy of RMB 20,000 (about $3,000) for a PHEV and RMB 50,000 (about $8,000) for a BEV. Changchun will offer RMB 40,000 (about $6,400) for a PHEV and RMB 45,000 (about $7,200) for a BEV. Shenzhen will offer RMB 30,000 (about $4,800) for a PHEV and RMB 60,000 (about $9,600) for a BEV. Hefei has not yet set individual rates but has set aside a budget of RMB 800 million (about $128 million) for subsidies. To qualify for the subsidies, there are minimum battery requirements (at least 15 kWh for a BEV and at least 10 kWh for a PHEV). As of 2014, fewer than 70,000 PEVs were on the road in China, far from the target set by the government of 500,000 by 2015 (Bloomberg News 2014).

[4] The six cities permitted to provide additional incentives are Beijing, Changchun, Hangzhou, Hefei, Shanghai, and Shenzhen.

REFERENCES

Beltramello, A. 2012. "Market Development for Green Cars." OECD Green Growth Papers. No. 2012-03. OECD Publishing, Paris.

Bloomberg News. 2014. "China to Exempt Electric Cars From 10% Purchase Tax." Bloomberg Business Week, July 9.

BP. 2013. "Statistical Review of World Energy." http://www.bp.com/en/global/corporate/about-bp/energy-economics.html. Accessed July 16, 2014.

Brand, C., J. Anable, and M. Tran. 2013. Accelerating the transformation to a low carbon passenger transport system: The role of car purchase taxes, feebates, road taxes, and scrappage incentives in the U.K. Transportation Research Part A: Policy and Practice 49:132-148.

Deshayes, P. 2011. "Electric Cars Take Off in Norway." Phys.org, May 10. http://phys.org/news/2011-05-electric-cars-norway.html.

Doyle, A., and N. Adomaitis. 2013. Norway shows the way with electric cars, but at what cost? Reuters, March 13. http://www.reuters.com/article/2013/03/13/us-cars-norway- idUSBRE92C0K020130313.

Dutch Ministry of Economic Affairs. 2014. Electromobility in the Netherlands: Highlights 2013. http://www.rvo.nl/sites/default/files/2014/04/Electromobility%20in%20the%20Netherlands%20Highlights%202013.pdf.

Ingram, A. 2013a. "Electric Cars: 12 Percent of All New-Car Sales In Norway Last Month." Green Car Reports, December 3. http://www.greencarreports.com/news/1088856_electric-cars-12-percent-of-all-new-car-sales-in-norway-last-month.

Ingram, A. 2013b. "How Much Is Norway Paying to Promote Electric Cars?" Green Car Reports, March 15. http://www.greencarreports.com/news/1082915_how-much-is-norway-paying-to-promote-electric-cars.

Ji, S., C. Cherry, M. Bechle, Y. Wu, and J. Marshall. 2012. Electric vehicles in China: Emissions and health impacts. Environmental Science and Technology 46(4): 2018-2024.

Jie, M., and Y. Hagiwara. 2013. "In Ghosn We Trust Tested as Nissan Electric Push Falters." Bloomberg, March 20. http://www.bloomberg.com/news/2013-03-20/in-ghosn-we-trust-tested-as-nissan-electric-push-falters.html.

METI (Ministry of Economy, Trade, and Industry). 2013. "Our Policy about Promoting EVs." Presentation to the Committee on Overcoming Barriers to Electric-Vehicle Deployment, Tokyo, Japan, December 9.

Nelson, T., and M. Tanabe. 2013. "Japan Continues to Offer Electric Vehicle Incentives." Dashboard Insights, September 12. http://www.foley.com/japan-continues-to-offer-electric-vehicle-incentives-09-12-2013/.

Nissan. 2012. "Amsterdam's TAXI-E Takes Zero Emission Push to the Streets." http://reports.nissan-global.com/EN/?p=3348.

Ouyang, M. 2009. "Development of EVs in China." Presentation at U.S.-China EV Forum, Beijing, China. September 28.

Tesla. 2013. "Electric Vehicle Incentives Around the World." http://www.teslamotors.com/incentives/japan. Accessed February 14, 2013.

Zeng, J. 2013. "Subsidies for Green Cars Not Bringing Expected Results." China Daily, April 1. http://www.chinadaily.com.cn/business/greenchina/2013-04/01/content_16364018.htm.